Letters are among the most significant memorials a person can leave behind them.

JOHANN WOLFGANG VON GOETHE
1749-1832

Dad Before We Knew Him

Letters Home From Aldro Lingard

1938–1944

Edited by His Daughter, Christine Lanoue

BOHANNON HALL PRESS

Printed in the United States of America
First Edition

All Photos © by Aldro Lingard and Friends
Cover Design by Bohannon Hall Press
(Cover Photo: Lt Col Lingard warming his hands on Mount Vesuvius)
Edited by Christine Lanoue

Library of Congress Control Number: 2025926519

Publisher's Cataloging-in-Publication Data

Lingard, Aldro, 1917-1984
Dad before we knew him : Letters home from Aldro Lingard 1938 – 1944 / by Aldro Immanuel Lingard ; edited by Christine Lanoue ; photography by Aldro Lingard and Friends
Niceville, FL: Bohannon Hall Press, 2025.
174 p. : illustrated- ; 23 cm.
First edition.

1. Personal Correspondence. 2. Military Life – 20th century 3. Title. II. Lanoue, Chtistine, 1947-
D811.L56 D3 2025 2025926519

ISBN 978-1-962995-14-6 (hardcover)

Published by Bohannon Hall Press

Table of Contents

PREFACE
Introduction to Dad's Letters

While clearing out everything in Mom and Dad's garage prior to selling their house, I was thrilled to discover a large box containing carbon copies of what appeared to be every letter Dad had ever written.

He'd always typed his letters with carbon copies, but I'd assumed that was because he sent copies to more than one person, hence the Dear Chris, Bob & Jan letters we received in college and beyond. But I'd never realized he kept copies for himself, much less that he had kept them through the many moves of an Air Force family – 15 for him and Mom that I can count.

More recently I made an equally exciting discovery: a box labelled "Grandma's things" that contained all the "Dear Folks" letters he'd sent her from 1937 through 1944. What a treasure! The letters came from his ROTC camp before and after graduation from college, then from Fort Monmouth in New Jersey, followed by flight schools, and on to Puerto Rico, where he met and married Mom before heading to North Africa and Italy during World War II.

What I have sought to do here is transcribe the parts of his letters that I felt would be of interest to his children and grandchildren, just as he wrote them. Any comments in italics are by me. If a passage started or ended a letter, I included "Dear Folks" or his closing 73, which is ham-speak for Best Regards; 88 means Love & Kisses. Dad was an avid ham radio operator, or "ham." Photography was also a lifelong passion often mentioned in his letters, and his cramped quarters didn't stop him from pursuing both activities, as shown in a photograph of his room.

I hope you will enjoy reading these letters as much as I have.

CHAPTER 1

Letters from ROTC Camp Custer

Summer 1937

Although Dad didn't graduate and move away from home until 1938, his first letters to his family were written in 1937, when he spent several weeks at an ROTC camp before his senior year at the University of Wisconsin, Madison. The camp was called Camp Custer and it was located in Battle Creek, Michigan.

June 25, 1937
Dear folks—

Well, the weather's been awfully changeable here; one downpour after another with hot spells between. Right now we're sitting in our tent listening to an undecided drizzle which may any time turn into a flood.

These uniforms are an awful nuisance. We have flannel shirts, and wearing a flannel shirt tight around my neck doesn't add anything to the efficiency for which the army says it's noted. My neck is sore and itching and so are my arms and back and it's hot too. Well, it's only five more weeks now.

Wednesday I was on KP. Wednesday afternoons here are regularly free, and it was just my luck to get K.P. on Wednesday and have to work 14 hours in the kitchen! One thing about that, though — for K.P. we wear denims instead of our regular uniforms. The overalls I've got don't even cover my ankles but they feel all right . . .

Waterman and I are arguing about the advantages of hand-written versus typed letters. He says a typewriter lacks a personal touch. I said I'd rather be able to read a letter than have to guess at the sweet nothings that might be in it – referring to a letter he got today which I couldn't read. He said that was an advantage.

7/4/37

Dear folks—

Well, there's not much doing today. We have tomorrow off too, and a lot of the fellows have gone home . . . Westerman went to Madison to see his girl. He telephoned her long distance last week to tell her he was coming. It so happened that Roth Schleck, who is also here, phoned his girl at the same time and the calls got mixed; Schleck talked to Westerman's girl and was pretty disgusted with the mixup. The funny part is though, that Westerman, who talked to Schleck's girl for 5 minutes, didn't realize that a mistake had been made and told the girl when to expect him and everything, thinking she was his girl. He didn't find out the mistake until a day later when Schleck told him! . . .

This week we've done pole climbing, a 9-mile practice march in one morning, radio and telephone communication, and we spent one day on chemical warfare. We had gas mask drill and then, with masks on, went into a chamber filled with tear gas. We couldn't tell the gas was there, but later they ran us through a strong tear gas concentration without masks and its effect was very noticeable. We could hardly see at all, and our whole faces burned for about ten minutes afterwards. The officer who was demonstrating said tear gas took people's minds off their worries. It does.

Pole climbing isn't as easy as it looks. The spurs will hold all right if you get them in right, but that's the trick. Going up isn't very hard, but coming down things seem pretty unsteady.

I'm getting used to shaving with cold water. It's supposed to be hot all the time but it isn't. Incidentally, we don't have to shave every day necessarily.

Wednesday, July 21

This would be a better picture, I think, if I hadn't known it was being taken and had tended to my potato peeling, but because I did know I faced the camera and took up a very nice pose, with the result that the picture isn't nearly as interesting as it would have been. The potatoes were in the two big cans beside me; I took them out of the nearest one, cut out the eyes – they'd already been run through the peeler – and then tossed them over into the other can. It's a nice way to pass the time. I happened to be out there alone at the time the picture was taken, although usually there are three or four KP's peeling.

Sunday 10 am. *(no date)*

"Well, we've started into the finishing stretch of our work here. We're winding up the camp work with a series of field problems – signal communications systems for a division including radio, telephone, and telegraph. Each school has one day when its students act as officers and take charge of the different phases of the work and the other schools work under them. They try to rotate us around among the different jobs – message center, construction, telephone switchboard operation, radio and telegraph, but there aren't enough of us who know code well enough and those of us who do will be on radio or telegraph all the time. I like it but I'd like other work too."

"Tomorrow afternoon I go on guard duty. After that I won't get any more KP or guard for the rest of camp. I've heard that guard is easier than KP."

June 1938

Grandma wrote in her Lingard history that Dad did not attend his graduation from the University of Wisconsin in 1938 because he'd gone to New York for some job interviews that didn't pan out. About graduation, she wrote, "We saw the engineers as they came in line to get their degrees – there were Paul Ketchum and after him Wayne Mitchell. Aldro should have been in his place between them. We choked a bit."

Dad spent the last two weeks of June at a camp in Michigan. His letters suggest that they were also written from Camp Custer.

6/13/38

Dear folks—

I've spent all day today at RCA; I saw Mr. Aiken at 10 in the morning, went around interviewing a lot of other men in the plant, and didn't get through till 5:30. . .

Last night Biggs and I were talking about Philadelphia and I mentioned the fact that they seemed to have a lot of alleys over there. He told me they weren't alleys, they were streets. They're all paved with bricks, too, and you know what a mess brick paving is. In eastern Ohio they use bricks to pave highways; I think I'd rather ride on a dirt road. . .

I plan to stay here tonight and go over into Philadelphia in the morning to see about a couple of possibilities there. After that I'll go up to New York, probably stopping on the way in Newark, and look at a couple of more places there. The two Philadelphia companies I want to see are Leeds & Northrup and the International Resistance Company; in Newark, the Weston Electrical Instrument Co. and American Transformer Co., and Bell Telephone Laboratories in New York City. After that I'm not sure where I'll go. It will probably be Wednesday night by that time, and time to think about heading West again.

6/16/38

Dear folks—

Everything came through OK – grades, two letters, letters from Taylor, I.R.E., and War department. There was a little delay in getting my trunk out here from Battle Creek, but it got here this morning and now I've got everything. . .

You'd better not send me any more clothes just yet, because I've got plenty for the next two weeks and what I'll need after that depends on what I do. About getting out of the army if I should want to, there's no written provision about that but it's been done over and over with no trouble. I'm sure I could get out if I should get a job, and if I don't get a job, I think it will be a good place for me to be for a while.

6/20/38

Dear folks—

Today we had our first regular day of duty. We had a lecture this morning and another this afternoon. It's been hot, and to add to it we're required to wear our uniform coats at mess. Right now I'm sitting in the tent with not much on except breeches, and the mosquitoes seem very much aware of the fact. . .

We went swimming in Lake Goguac just outside Battle Creek yesterday afternoon.

6/22/38

This morning we were out and watched a demonstration of an airplane attack on a marching column of infantry. The airplane, flying low, would come up from behind spraying machine gun fire on the column. As soon as the plane appeared, the column would scatter and everyone would shoot at the plane. Riflemen are able to get off about two well-aimed shots during the time the plane is in range, and the machine guns in the column would just stay right on the plane. It took about 7 seconds to unlimber a machine gun from its cart and open fire on the plane, leaving about 10 to 15 seconds for firing. Everything was demonstrated with blank cartridges, of course. After that they demonstrated going into camp for the night (bivouac), and concealment from aircraft.

Tomorrow night we're to go out after dark and go through an exercise in following a compass course through unknown territory at night. Flashlights will be limited to use in locating positions on maps; we won't be able to use them to find our way around, the idea of course being that in actual warfare we wouldn't want to risk being seen by the enemy. We've been told to wear fatigue uniforms (overalls) and old shoes. Most of us are borrowing service shoes from the ROTC's.

So far we haven't had a thing pertaining to Signal Corps work; it's all been infantry stuff. Tomorrow morning will be spent in Signal Corps work, and that's all of that we will get. . .

This morning I caught a Michigan mosquito I was going to send you, but it got worn out in my pocket. Without seeing them, you wouldn't believe they could be as big as they are.

6/24/38

Dear folks—

We're sitting in our tents this afternoon waiting for it to stop raining. We went out on a tactical walk right after dinner, and as soon as we got where we were going it started to rain so we had to come back. A few strokes of lightning came pretty close, and one group of us got a little shock in the feet from one stroke. That's the closest I've ever been, although I don't know where the bolt actually hit.

We were out last night on a night compass course. We were taken in trucks out about two miles from camp and told to follow a certain compass course across country until we hit camp. The course went through a woods with thick underbrush and a swamp which we had to go around, and of course the mosquitoes were pretty bad. Our group – three of us – after going about 3000 yards by compass in that kind of territory, finally ended up about 20 yards from the spot we were supposed to have hit.". . .

I still don't know anything about coming home after camp, except that the more time goes by with no information the more certain it seems that that will be the case. My army appointment is supposed to start the first of July, but it will probably delay about a week.

73 -

Aldro Lingard

6/27/38

Today we had a chemical warfare demonstration, pretty much like the one we had last year except that this time they didn't give us tear gas without masks. We did get it with masks, though, and although it didn't bother us with masks on, as soon as we came out of the gas chamber and took off our masks the gas which had worked into our clothes went to work on us and the tears flowed pretty freely.

CHAPTER 2

Letters from Fort Monmouth, New Jersey

July 1938 – June 1939

In his "Childhood Memories from Mount Horeb" Dad wrote: "When I graduated from college in 1938, only four members of my graduating class in electrical engineering had jobs promised by graduation day. . . I was one of two other members (new ROTC graduates) who were selected for a year's active duty with the Army Signal Corps. At the end of that time there still weren't any jobs, but again I was one of a few successful applicants for Army flying training."

In July he began his year with the Signal Corps in Fort Monmouth, New Jersey. This was his first exposure to ocean water and to New York City. The July 16 letter provided detailed instructions on how to ride a wave and the Nov 30 letter included impressions of the big city.

1938 saw the birth of the chain letter with Carl Walter and his other college buddies as described in the Sept 5 letter. This chain letter continued for decades.

7/7/38

I got here yesterday about 2 o'clock, and haven't done much since then except take a medical exam and get settled down. We have a two-room apartment here; a bedroom, tile bathroom with shower, and a closet, and a study room or office.

We're about 3 miles in from the ocean, and 37 miles from New York, I think. . .

The address here is just Fort Monmouth, New Jersey.

73

Aldro Lingard

7/11/38

For supper tonight we had pork chops with paper handles on them so we wouldn't get our fingers greasy. I've never seen those before. The service over there is pretty fancy all around; we have 3 spoons, 2 forks, and a knife at every meal, our butter is in silver dishes, and bread is on silver trays. After supper we played ping-pong a while on the porch of the Officers' Club, after which another ham and I went to call on an officer here whom we met at camp last summer and who is a ham. He wasn't home, so we dropped in on one of the majors. He was telling about all the historic spots around here and how his battalion had been called out to help keep order at the time of the Hindenburg crash.

My room-mate is Miller, from Green Bay. He was out playing golf this afternoon, and it seems he happened to hook up with a general out there. He's just been telling about the General's golf. . .

I haven't been swimming yet – naturally, without a suit – neither has anyone else. The nearest place is the ocean, about 3 or 4 miles away, and I guess it's pretty hard to get there by bus – that is to the place we want to go to. The officers here have a place kind of to themselves. It's fairly easy to get to the Long Branch beach – that's a town (20,000) about 3 miles south of us – but according to a couple of the fellows who've seen the place, it's not very nice there.

7/3/38 *(This letter must have been misdated; probably written 7/13)*

Dear folks—

Miller and I got this typewriter just a couple of days ago together/ It's about like mine at home, except I think it's about a year earlier model/ It doesn't have some things that it ought to have – for instance, a backspace, and one thing it has that it definitely shouldn't have is a shift for the period. You notice I ended my first two sentences with a / That's what's where the period should be. In order to hit the period you have to shift. / shift … I'll probably get used to it/ -- or will I?

We went swimming this afternoon/. The tide was way in and the drop-off into neck-deep water was pretty sharp. The waves didn't break until they got just about in to the drop, and if you tried to ride them in from there you'd get dumped way up on the beach with a swimming suit full of gravel you'd picked up from being pushed along the bottom. But we got cooled off anyway.

7/15/38

Went swimming this afternoon. It was awfully calm today. I've never seen the water as smooth as it was. It seems the ocean is never the same two days, and when it's rough then it's fun and when it's smooth then it's just like Lake Mendota, except that it's salty, and you have to be careful about opening your mouth.

7/16/38 *(how to ride a wave)*

Went swimming at Seabright this afternoon. The waves were coming in fairly well, a three-footer occasionally. . . It's lots of fun to ride in on the advancing wave fronts; watch for a big one to come along, get close to where you think it's going to break, and just before it does, give a big push in toward shore. Then the wave will hit, and carry you in 75 or 100 feet before you know what's happened. If you're good, you can ride in without getting your hair wet, but it's more common to get flooded. . .

We start on our regular school schedule tomorrow afternoon. It will last six weeks; after that I don't know what we'll be doing. The first week we'll go into Signal Corps organization, combat orders, training methods, message centers, military cryptography, and pigeons. Cryptography is the job of putting messages in such a form that they can't be read by anyone who doesn't have the key, a job which is handled by the Signal Corps. Carrier pigeons are used by the Signal Corps too, but only a little.

7/20/38

We've had a lot of sea food around here. Tonight for the first time I took some steamed clams. I don't think they're worth while, although there's nothing wrong with them. So far I haven't got up nerve enough to eat a lobster, but I will.

7/21/38

Played three sets of tennis this morning, and lost all three. Last night we'd been down to Asbury Park, about fifteen miles south of here on the ocean. It's a big amusement place. This afternoon we were swimming, and yesterday afternoon too. The waves were pretty big today, and they were breaking in close to shore too. The tide was in. Every once in a while, a wave would hit while I wasn't watching, and then I'd pick myself up from shore with most of the skin scraped off my chest where I'd scraped bottom on the way in, but it was fun anyway. Thursday the waves were so high that very few people went in at all. I did and the waves came very close to tearing my trunks off. That's' an embarrassing position, because they'd not only undress you but pick you up and throw you up on shore too instead of leaving you in the water where you'd be covered.

Tomorrow, we climb poles. Maybe I'll learn how this time.

7/23/38

Dear folks—

Just came from the theater – "Exiled to Shanghai." I happened to go because I got in free tonight, and it turned out to be pretty good. I got in free because I'm Officer of the Day today, or rather, assistant officer of the day. They're giving each of us one turn at assistant before putting us on the job alone. It means being in charge of the guard and looking after the post in general. I've got to stay up until midnight and then go out and inspect the patrols. . .

We've been having a drought today. This afternoon it didn't rain for two hours. . .

Sunday night – 9:45

Alan was here this afternoon, also Dunning, from Minnesota, whom I met at camp last year and who's working with him in New York. We went for a walk around the post, dropped in at the guardhouse to look at the radio system, and after supper went to the show. It was "Her Jungle Love" – on the romantic side. . .

We won't be able to get away next Saturday, but the following Saturday we're going to try to get up to New York. We can get there by government boat and I won't cost us anything; by train the round-trip fare is $1.90.

It hasn't rained all day. I don't know what to think.

7/28/38 *(phosphorescent sea water in New Jersey)*

Everyone around here wants to get cameras like the one I got. . . Mautz (California) has an enlarger, and quite a few of us are interested in fixing up a dark room of some kind. Youngren has been sitting here in my window playing ukelele chords on my guitar.

No one can say we're not a forward-looking bunch of people here. Everyone is making plans for going home Christmas, and already a snow-ball fight between the Northerners and Southerners is scheduled. We've got two fellows from California, two from Texas, one from Alabama, and one from Florida. Florida says he's seen snow once, but the others don't say anything. . .

Friday night the bachelor officers threw a party in the hall outside here. They set up two radios, turned on all the way, and danced all over the place. It was pretty noisy and I didn't like it very well. About 12:30 a bunch of us went swimming at Seabright with one of our instructors, and that was fun. It was awfully cold on the beach, but the water was nice. It was low tide, and shallow a long way out. There were a lot of crabs around too; every time I touched bottom it seemed I stepped on a crab, so most of the time I didn't touch bottom. Swimming made the water phosphorescent; everywhere I went the water glowed behind me. I don't know just why that is. There seems to be phosphorus in sea water, and agitating the water makes it glow.

We've been having field wire work this last week, and you should have seen our uniforms Friday afternoon after we were all through climbing trees in them, laying wire lines, installing switchboards, and getting thoroughly dirty in general. Tomorrow, we start something different. We go into the enlisted men's school and do the work they do.

7/28/38

It hasn't rained for several days now, but it's very damp. My tennis shoes got wet a couple of days ago and they aren't dry yet.

11 Aug

I'm off on a letter-writing session this afternoon/ I'm Officer of the Day again – that comes round about once a month or a little oftener – so I've got to be here but haven't anything much special to actually do. One nice thing about it is that I don't have to wear a coat to supper. Just cotton shirt and pants – regular day uniform– the only difference is that I wear a saber and Sam Browne belt. I get in free to tonight's show, too – saving 14 cents, which I probably wouldn't have spent anyway.

9/5/38 *(the birth of the chain letter)*

I got a chain letter a few days ago from Alan – Wayne started it and it's to go the rounds of our graduating EE class. It's an interesting idea. Everyone writes a letter and puts it in with the others and sends it on, and when it comes round to him again, he takes his letter out and puts in a fresh one, and so on. I didn't hear much new in the letter – Wayne and Dick Nelson and Alan were all that were in when it got to me, and I knew what all of them were doing, but next time around it ought to be interesting.

I'm sending a few pictures of me. One was taken one day when we were out in the field all day operating radio sets. Another is me and my camera.

9/11/38

The social season here starts in two weeks, which is going to be a mess. Officers aren't invited to things, they're ordered. Only ladies are invited. We've got to wear white gloves too. I don't know if it will be that way all year—this thing in two weeks is a reception for the new commandant, and they probably won't be so fussy after that. Orders for the reception were issued Friday, which was two weeks ahead of time so no one can have a valid excuse for not being there.

9/17/38 *(first mention of war)*

I haven't been reading the papers much, and your letter was the first I'd heard about any serious war threat. From what I've heard, though, the people in the middle west seem more worried about it than those here. Incidentally, the Signal Corps doesn't use flags any more; they use telephone, telegraph, radio, messengers (mounted, motorcycle, and foot), pigeons, panels for communication with airplanes without radio, and pyrotechnics (fireworks) sometimes. They are working on teletype, too, and they'll probably be using that soon.

73

Aldro Lingard

9/23/38 *(I have this clarinet if anyone wants it. $30 in 1938 is about $660 now)*

Dear folks—

I think you might as well sell my clarinet if you get a good chance, as this one seems to be. $30 I think is a very fair price both ways. As far as getting it fixed up is concerned, there's nothing wrong with it except that it needs to be played, and of course it needs a new reed. That will cost from a quarter up to 75 cents or so. The mouthpiece has a rubber pad on it for the teeth to rest on without biting; that's not standard, but it can be taken off if they don't want it. Also, if they'd rather rent it for a while to see if the girl can play it, that would be all right too. I'll have to leave all that up to you. It seems best though that it be used, and I probably never will. . .

More tomorrow—time to eat now.

73

Aldro Lingard

9/26/38

Dear folks—

We started a new course again today. The last four weeks we've been on radio circuits, sets, and operating procedure. The next four weeks we spend studying telephone circuits. The course is pretty simple. It was designed for raw enlisted men with no technical background, and they put us in there because they haven't anything else to do with us until the battalion gets back from maneuvers, which will be about the end of October.

10/2/38

We went up to New York yesterday on the boat—rode up to Fort Hancock with one of the regular officers and his wife. It was a pretty windy trip . . .One of the things I got was some stuff for developing my own films, and a tank. Maybe now I can get some decent work in a reasonable length of time, which none of the professionals around here seem to do. . .

Peanut wagons and bootblacks are the two most common things in that town. I counted eighteen bootblacks on one side of the street in one block. And they do business too. I guess people in the East don't shine their own shoes.

10/11/38

Saturday one of the regular officers called up and asked if anyone wanted to go to New York. . .

When I got home here I found the ghosts around here had been loose. My bedspring was up on top of the bookcase, one end of the frame was in the shower, and the other end in the closet, and the bedclothes were just dumped on the floor. Well, that wasn't very convenient, but I put it back together and went to bed right away. It seems one of the fellows had found one of his cigarettes had been mistreated and suspected me. He was right too. It was fun anyway. There's been a lot of horseplay like that lately. Last night the same fellow—Linn, W9NZF, Chicago, had been sitting in on a coffee session upstairs and had fallen asleep. He's an awfully sound sleeper, and the fellows poured cold water on him to wake him, but it didn't. It took about fifteen minutes to rouse him. . .

I've just finished the four weeks' telephone course in two weeks and two days, and I'm starting to work on teletype now. That's very interesting, and it's all new to me. I've never had anything on it before. What it is, is a couple of machines something like typewriters, connected together electrically and whatever is typed on one is reproduced on the other. Western Union is using teletype more and more, and it's replacing the old style of telegraphy very fast.

10/16/38

We were having a bull session one night a few days ago with Spencer and Davis, from Florida and Texas respectively. Spencer was telling about one night he'd been out on a date and someone had fixed his car so it wouldn't run. He'd found the trouble, but first he'd run down the battery trying to start the car…it seems it was awfully cold too, he said he almost froze to death. I asked him how cold it was, and he said it was down to thirty. Thirty above or below? Why, where he comes from, they don't have only one side of zero. It's a lot of fun listening to those Southerners talking about freezing when it gets down to forty, and then telling them about our thirty-below spells. They just haven't any use for weather like that. Miller had his ice skates sent up here a few days ago, and Texas was very much interested. Spencer and Davis spent about an hour putting on Miller's ice skates and walking around in them, and then the whole gang of us went roller skating down on the other side of Long Branch. Roller skating was interesting too. I don't remember when was the last time I had them on, and my feet don't seem to remember either.

10/24/38

We went to New York Saturday on the boat. Met Walter at Fort Hancock and he went up with us, and then we met Alan at Western Union. We had lunch at the Western Union cafeteria, and after that he showed us around the place. Most of their traffic is handled by teletype now, with a few manual operators. They have messenger girls running around all over the place, and they're all on roller skates. Very interesting. After that we went up to Forty-second and Broadway and looked for a good show, but didn't find one so we went to the RCA Museum of Science and Industry for a while. We were there for a couple of hours I guess, and by that time it was time to go home. We all went back on the boat, and Walter had brought his car to Fort Hancock so from there we came back with him. Sunday morning we looked around a little, went through the pigeon loft here, and then Alan and I played tennis and Walter, Miller and a captain went out to play golf. In the afternoon we went down to see the naval air station at Lakehurst, where they have all the dirigibles and blimps and things. The only dirigible this country has now is the Los Angeles, but they've got a lot of small blimps, and they're at Lakehurst. The Los Angeles isn't such an impressive looking bag, although it is big, of course. The Akron and Shenandoah were much bigger, before they crashed. They've got a few pieces that were left from the Hindenburg on exhibit in the hanger too.

10/26/38

Dear folks—

Wednesday night. I thought I'd change my letter-writing schedule, because there's not much sense in writing just a couple of days before I get a letter from you and then waiting a week before answering that. . .

Linn and Davis across the hall have been building phonographs lately. Linn went to a show with me one night – it was "Cocoanut Grove"—and liked the orchestra in it so well he decided to build himself a phonograph so he could play their records. So he did, and Davis got interested too and now they're building one for him. Every once in a while, I hear the thing blasting away with the volume control up full. I'm going to rebuild my transmitter. . .

They've got a big weekend planned here this weekend. Friday night is a Hallowe'en party; Saturday is the Army-Notre-Dame game in New York, and for those who don't go there'll be a party and broadcast at the Officers' Club, and then a football supper and after that a dance. I think I'll go to New York and see Alan again. . .

I've been working on page printer teletypes this week. Last week it was tape printers – the machine prints the incoming message out on a long paper tape. On the page printer, the message is printed on a long strip of paper, page width. Lots more gadgets. Yesterday we put a few assorted short circuits and open circuits on a telephone Reed was supposed to be fixing, and the troubles were so well placed and so well hidden that it took him all today to find them. So late this afternoon he went to work on my teletype and fixed that up. He was waiting for me to put the machine together and find it wouldn't work, but I found the trouble – it was very simple – without his knowing about it. So, there I was, and I turned on the machine and it worked perfectly, and there was Reed wondering what you have to do to one of those teletypes to keep it from working anyway. Tomorrow we'll probably fix his telephone again. That's about all we've got to do these days. More fun.

73

Aldro Lingard

11/2/38

Dear folks—

Yesterday we started our new jobs with the companies. I'm assigned to the 1st Radio Intelligence Company, and my job right now is that of assistant supply officer. I've got practically nothing at all to do—in the two days I've been there the work I've done has been to move four radio direction finders (with a truck) from one building to another and unpack about three chests of equipment and throw the stuff on shelves. . .

The job of a radio intelligence company in wartime is to listen in on as much enemy communication (radio) as they can pick up and make use of the information so obtained, and to locate the enemy stations by direction finder stations. We may get a chance to work with some of the sets they have there, but not very soon. I'll be assistant supply officer for a while, and then probably assistant motor transport officer (we haven't any motor transportation) and then probably assistant mess officer. We haven't any mess either; our company messes with another company.

11/9/38

I don't remember whether I ever said I like the place here or not; well, everything is very comfortable and easy and there isn't a thing to be dissatisfied with except that it's too easy. . . It's just the way things are in the company; there's so much work to do, and exactly three times as many officers as are needed to do it, so what can we do but sit around most of the time? Well, we've got a nice place to live, we get good food most of the time, although we pay for it, we work about a day's worth in a week's time and draw our pay for it. Very nice, although not particularly satisfying. I know I'll never have things as easy as they are right now after I get out of here.

11/16/38

Dear folks—

We had this afternoon off so we worked on our transmitting antenna. I've been up in an evergreen tree most of the afternoon, Linn's been up on the roof, and Davis and Dumas and Christof and a few other people and Backus were on the ground doing things. We're not through yet – it's a kind of complicated process because the antenna has to go above a power line and in putting it up we have to be careful not to let it come in contact with the line. We'll finish it tomorrow, I think. We had to boil some wooden spacers in paraffin – we did that over in the kitchen of the Officers' club, and I guess the cook over there is about ready not to be surprised at anything we come around wanting from her. We're always over there getting baking powder cans or some kind of tin cans or jars or all kinds of junk from her. She doesn't mind though. Today I gave her a picture I'd taken of her cat, and she gave us a couple of apples. They were good too. . .

It seems everyone around here is either a photography enthusiast or a ham or both. The two seem to go together for some reason. The signal Corps, which handles all the army's communication equipment, also does all the photography for the army.

11/22/38

There's not much doing over at the company; all the time lately when I have been over there we've spent most of the time arguing and talking with the captain about vector analysis – methods of solving complex electrical circuits and problems. It's the first time I ever ran across an army officer who remembered anything of what he was supposed to have known about that stuff at one time. It all grew out of a discussion between the captain and one of the other officers on whether it was possible to teach radio antenna theory to enlisted men – we're going to be doing that before very long – without dragging in ideas that they don't know anything about, and I got into the argument and now we're all in it. . .

Thursday last week we put up our transmitting antenna. We had to take it down again, though, because it was too close to a 2300-volt power line. But we put it up again, differently, and now we've got to get approval for the new installation. And we've got to make some changes in that too yet. I think we may get on the air very shortly, though, because if some stuff I ordered comes tomorrow I should be able to get my outfit on the air by this weekend. Last Sunday morning

I heard W9RNX, Russ Morris, on the east side in Madison. He was awfully hard to pull through, though; there were about six other stations almost exactly on his frequency. Linn is building a transmitter too, and his is just as close to being done as mine is.

11/25/38

Dear folks—

Thanks a lot for the stuff; the box came yesterday morning about 9 o'clock. I don't know how it happened – I didn't think there'd be any mail delivery on Thanksgiving. Anyway, the brownies are good, and the candy unusually good, and I ate the carrot right away and it wasn't dried up.

Aunt Inger sent me some lefsa, and that came at the same time. I took it over to the club for dinner – there were five of us at our table. Youngren was a Swede from Minneapolis, and of course he knew what it was; Miller will always try anything, and he liked it pretty well. Edwards (Alabama) and Andrix (Ohio) wouldn't touch it. They said they'd outgrown the age when they liked to chew on dishrags. We had a lot of fun. The dinner itself was really something too. We had a whole turkey between the five of us, and I don't know when I've ever eaten so much before.

11/30/38

Went up to New York last Saturday . . . One thing I did--I happened to get into Macy's department store. Never saw such a place. I don't know how many floors they have--I rode the escalators up to seven and then back down to the basement and by that time I was tired of escalators so I walked up to the main floor. But every floor -- area a complete big city block -- was jammed so full of people you could hardly move around. A lot of New York is that way--the radio store district on Cortlandt Street, Broadway near Times Square-- it's impossible for two people to walk abreast; they have to zig-zag through holes that open up in the crowds, and have to do it in a hurry before the holes close up again.

We had quite a bit of snow last weekend. Friday it snowed, and Saturday night more came; Sunday morning there was about 8 inches of soft, packing snow, and Miller and Waterman and I went out and threw a few snowballs around and then went for a walk out to West Gate--a couple of miles. We came back cross-country through the woods; we chased a squirrel, and he fell out of a tree and nearly went down Waterman's neck . . . Waterman, being from California, isn't used to snow, and he got a real kick out of it. I took a few pictures of him rolling around in the snow in a sailor suit. The thing that's really interesting, though, is Edwards. He's snow removal officer for the post, and being from Alabama he's never seen any snow before. We went out to where he was in charge of the snow removal platoon on the first day and heckled him. We had more fun--he was just standing around looking bewildered while the men shoveled snow.

12/7/38

Linn and Davis always have trouble getting up in the morning; you could load their beds on a truck and haul them away and they wouldn't know a thing about it. So they took an alarm clock and mounted the works in a tin can, and inside the can they put a transformer and a doorbell, so arranged that when the alarm goes off, instead of just ringing it closes a contact and starts the doorbell hammer beating against the bottom of the can, at the same time turning on a light so they can see the face of the clock to see what time it is. You can hear the thing all over the building when it goes off, and neither of them has overslept when he had that gadget to wake him up.

12/14/38

Dear folks—

Been working on the transmitter most of the afternoon, except for an hour or so when I went to Red Bank. No new dope on coming home . . .My leave is 12 days, which means I'll have to leave probably early in the morning of the second, or possibly sooner. . .

I'm sending a couple of moonlight pictures I took one night last week when the moon was so bright Youngren came in all out of breath and wanted to rush out at ten o'clock to take pictures. Well, it was a good idea. We took a few shots of Russell Hall -- that's Post Headquarters -- and then went over to take pictures of the cannon. We fussed around there for so long the guard came over swinging a big club to ask us what we were doing. When he got up close he recognized us, I guess. I don't know what it is about an officer, but the enlisted men seem able to recognize one no matter what kind of civilian rags he's wearing. We like to think it's some quality of perfection in bearing, of course, but we know perfectly well it isn't. . .

Incidentally, we got our radio station license at last. Our call is W2LTS.

More late-at-night monkeying around has been going on around here lately. Of course, Davis and Linn and I are always working on our transmitters and holding bull sessions at the same time. Last night Mautz was printing pictures. About ten o'clock he decided he didn't like the way his camera was working, so he loaded the thing up, took about ten pictures -- some of which, by the way, were the craziest shots -- and then asked me to develop the thing for him because he wanted to go on with his developing prints. So I did -- that was ten o'clock. It took an hour to develop, and that was eleven, and then it had to be washed a while and then hung up to dry, and that was eleven-thirty–no, twelve, before it got dry. And then he went ahead and made prints from that. I don't know what time he got to bed. . .We don't have anything much to do in the daytime, so we've got plenty of energy left at night so we stay up and play around. At home I never stayed up after ten-thirty without a very good reason; here I never go to bed before ten-thirty without a very special reason.

73

Aldro Lingard

1/4/39

Dear folks—

Well, I made it, but what a trip. According to schedule, I had 25 minutes to make my connection to get to Fort Monmouth. Well, everything went fine until just before we hit Cleveland; then the battery on the bus went dead and we lost an hour waiting for a replacement. We had a pretty good driver, though, and he managed to pick up half an hour of that lost time before the next driver took over. But that next driver was no good. He lost all of that half hour and an hour more besides just loafing along, spending too much time at rest stops and lunch stops. Well, it would be impossible, of course, for me to get that bus out of New York, and it was the last bus out of there too. So I got off the bus in Newark intending to take a train down here. They have a 3:30 train in the morning from Newark, but when I got to the depot, it was closed. They do things like that in the East. They may have trains running, but that's no reason why the station should be open. I didn't know what I should do then, and the time was getting short, so I asked a man standing on a corner if there were any other stations around, and he turned out to be a deaf mute. Well, I finally made it. I took a local bus to Elizabeth and got the train there and got in here at a quarter to six in the morning.

1/18/39

I'm glad to hear the waffle iron seems to be working better than when I was home. If it doesn't work right, don't put up with it. Send it back here. . .

I guess the army must have a pretty well organized publicity system. It's not particularly unusual for the newspapers to have the dope on what officers are going to do before the officers know themselves. That happened when Captain Kunesh was ordered to Hawaii from the University.

The mess officer job is interesting –it's the only interesting job I have. I ate over at the company this noon/ They had steak and mushrooms, and I knew there wouldn't be anything like that at the club so I stayed for dinner.

1/25/39

Dear folks—

We've been having real weather lately. First it froze and we had skating. Then it snowed and covered the ice, so they had it cleaned off and we went skating for one day – Saturday. Sunday it thawed and Monday it froze again and then the skating was perfect. Yesterday it thawed again and today it's frozen again – it goes in cycles, just often enough to keep the ice from getting cut up too badly. After the wind goes down it ought to be good skating tonight.

Went down to Long Branch a while ago to get our skates which we'd had sharpened. On the way back we came along the ocean shore for a ways hoping to get some pictures of waves, but the wind was offshore and it almost had the waves going backwards. . .

Margaret shouldn't study so much. I wonder what she'd do if she were taking engineering. . .

I finally got around to fixing up that picture of Davis. I'd draped him in a sheet, you know, and given him a red bulb in a socket and a book and stood him up on the table in our room and taken his picture that way. Well, I printed that and cut out the figure of the "Statue" and then glued that onto another picture I'd made of some sky. Then I rephotographed the whole thing on the copying camera over at the photo labs and printed it that way. . . About a picture of myself in uniform, by the way, I'll get one taken, but I can't do it now because I just had about the worst haircut I've ever seen on anyone.

Last night we stopped in a bakery in Long Branch and bought an apple pie with the idea that when the waiters came, brought in the usual mess with stuff on it for dessert, we'd go get our pie, unwrap it with much noise and comments on the dessert they were trying to feed us, and the fact that this time we brought our own, all while the Hostess looked on from her table over in the corner. It didn't work, though. The dessert was pie last night. It was chocolate pie, but pie nevertheless. We've had some good meals lately, but not what we should have for $40 a month. A couple of nights ago we really got the Hostess so mad she was about ready to boil over. We got kind of jolly at our table and started a little custom of throwing a salt-shaker at anyone that made a pun. Well, with salt-shakers flying through the air in all directions, she got kind of dissatisfied. Then Miller dumped a little pepper in his hand and blew it at Youngren. And Linn didn't like that, so he took another pepper shaker and took its cap off and chased Miller out of the mess hall with it, and all the time the salt-shaker barrage was going on. We're getting to be a regular bunch of ruffians. Well, if they want us to behave like gentlemen, they should feed us like gentlemen.

73

Aldro Lingard

1/31/39

Friday, I was officer of the guard. That's a new dirt detail post headquarters has thought up for reserve officers. The way it works, the officer of the day does just what he always did. The sergeant of the guard does just what he always did, and the officer of the guard acts as stooge for both. All he has to do is sit in the guardhouse for 24 hours. But it's a mean and useless job. They just found something for him to do now, though. Between 11:00 and 12:00 pm, he goes around through all the barracks and makes sure that enough windows are open in the squadrooms to give adequate ventilation to everyone sleeping there. What he ought to have is a nurse-maid's uniform.

B company is putting in a ham station. They're in the same barracks with our company. They haven't any officers who are hams, but that doesn't make any difference. They just turn the station over to a reserve officer who doesn't know anything about radio except what he's learned in the army and expect him to design and build it. Hastings is the one, so he's had Linn and me and Dumas all helping him unofficially; I don't mind -- it's interesting and I've got plenty of time anyway.

I got a letter from Professor Watson yesterday telling me about a job that may open up in Kokomo, Indiana and suggesting that I apply. I did – airmail. Sounds like just what I want to do.

We're going up to New York tomorrow afternoon for things and Linn and I will spend part of the time shopping around seeing what kind of prices we can get on parts for B company's transmitter. I don't know why we want to bother doing that. We asked the captain if it would be all right if we tried to get a little better prices on parts than catalogs listed -- if they'd let us go ahead we could cut the price by at least 15% with no sacrifice in quality of parts -- but it seems that Army Regulations has a process like that so tied up in red tape that it's impossible to do much of anything. It all boils down to the fact that we're trying to do the government a favor and the government doesn't want us to. You haven't any idea what Army Regulations are like; over at the office we've got a stack two feet high, and they cover only a few subjects.

The mess over at the company has been going fine, but there's trouble ahead. You see, we have two companies eating in one mess, which is run by B company. Well, B company has a mess sergeant who's no good at all. And the RI company -- that's my company -- has a mess sergeant who's one of the best in the army. Just lately B company's Sergeant went away for a while so ours took over and has been putting out really fine chow. B company's captain would like to keep our sergeant in charge, and all the men in both companies are definitely in favor of it. But our company commander isn't so sure. He has an idea he may have to take our mess sergeant away from the mess and teach him to drive a truck instead of letting him run the mess. Well that sergeant has been driving a truck into town every day for I don't know how many years, and he's probably one of the best truck drivers on the post, but there's the possibility he may have to turn the mess over to the other sergeant, thereby ruining the morale of two companies, so he can be taught to drive a truck. And the officer second in command of our company can't understand why I'm mad about it.

2/15/39

Just saw "You Can't Take It With You," and I don't know when I've seen such a good show. It's a comedy, but there's more to it than that. . .

The family in the show reminded me of the Thomason officers here – or rather of the small group we call Gadgets Incorporated. Three rooms here, right next to each other, are all filled with radio gadgets, phonograph gadgets, photography gadgets, and stuff in general. A few days ago a letter came here addressed to Gadgets, Incorporated, Building 271, Fort Monmouth – it was advertising gadgets. It was from Allied Radio Company in Chicago. In the show they had a pet crow, though, and that's one thing we haven't got.

3/1/39 *(after some comments about needing copies of his diploma and birth certificate)*

This whole mess of red tape is in connection with exams for permanent commissions in the regular army -- you know two of the 17 of us will be permanently commissioned at the end of the year, and these exams we're getting now are to pick the two. Well, I don't want any such job but I've got to take the exams anyway, and I needed the diploma to get out of part of it and I'll need the birth certificate before I can take the rest.

3/9/39
Dear folks—

I'm gradually picking up work to do around here. Now I've got to write a training regulation for using a new type of radio direction finder set we've got in the company. We've had the sets out in the field in snow, rain, sleet, and wind, and use them inside when the weather's been nice, and we've experimented with them ourselves -- the officers I mean -- and now we're supposed to know enough about them to write training regulations (meaning instructions) for the use of the sets. We're the only outfit in the army that uses these sets and we're kind of guinea pigs.

We're in a message center class now. The instructor doesn't know anything about message centers and we've all had the course four times before and worked with message centers, so it works out that we're teaching him the course. It's all right though; we've learned as much teaching him about message centers as we would having him teach us.

3/14/39 *(an evening in New York City)*

We had a lot of fun up there -- saw a show, got lost on the subways again, got stuff for Gadgets, Inc., which incidentally has a new member, looked all over Thirty-third street for a drug store we afterward found was on Forty-second Street, walked a mile to a photography store and found we were just around the corner from where we started, and finally got home about 11:30. We were tired, of course, but the next day we felt fine. . .

I just found a picture you might like . . .Linn in my room operating Dumas's transmitter (center) with his transmitter on the left above the table and mine at the right.

3/22/39 (*pastimes)*

I'm building a new enlarger; this one is going to have its own lens instead of using the camera as part of it, and it's going to be lighter and better all around . . .

I finally broke down and bought some new tennis balls this afternoon; we'd been playing all winter with some old dead ones of Mautz's.

Someone lately dug up a very dilapidated set of balls and an even more dilapidated set of cues for the still more dilapidated billiard table that's been standing out in the hall ever since we've been here, and that's been getting a lot of attention. I just finished a couple of games and now Edwards is out there playing by himself. The table is jammed in so that lots of times we don't have room for the cue on account of the wall, so for those shots we use the end of a baseball bat, and when that's too long we use my hammer. Some fun. I find I can usually get better accuracy by closing my eyes and swinging the baseball bat broadside than I can with a cue.

3/~~28~~ ~~29~~ Wednesday *(the dates were crossed out on the letters)*

Well, by the end of next week we'll be all through with this school business and I can go back to giving my full time to being assistant mess officer of a company that hasn't got a mess, and things. Another day or two and I'll have the training regulation I'm writing all in shape except for cutting stencils and things, which I don't do. . .

I had the enlarger set up last night for the first time – the new one, I mean. It seems to work pretty well – I had my 1 x 1 ½" negatives blown up to about 8 x 10 and they looked good.

4/5/39

Waterman and I were out and played five sets of tennis this afternoon. I think that's more tennis than I've ever played in one day before and I'm kind of tired now. . .

Griffin just dropped in and we were talking about my direction finder service test. I'm going to set the thing up Monday morning and spend a few days playing around and getting used to it before I start on the tests. They'll probably be conducted at some location several miles from the post. I'm going to be working at all hours of the day and night, too, I think, because direction finding conditions change at different times during the day, especially for long distance work. I was out to the Signal Corps Labs yesterday afternoon for a while, and they tell me I can have one of their engineers to help me any time I want him. This is going to be fun.

4/12/39

I started working with the direction finder Monday; we had the tower set up near the company area for two days, and this morning we moved it out into the country about 5 miles, to the position we're going to use for our tests. The tower was kind of awkward to load on a truck without taking it apart, but we made it. Tomorrow morning we'll start with our tests. We got the tower set up in the field today and found that it wasn't all through being built yet -- it's a wind tower about 15 feet high, and it fastens to a wooden base with bolts, and the outfit that built it hadn't drilled all the holes for the bolts, so I guess I've got to take a bit and brace out tomorrow morning with me and have one of the men do that in his spare time. I'll probably take a transmitter myself in a truck and cruise around the country and let them take bearings on me, and then compare what they got with the correct bearings on a map. That will be the way all our tests will be made, only they'll all be made under different conditions.

Sunday and Monday Lynn and I had a couple of "Walkie Talkies" out that we borrowed from the school. They're little portable transmitters and receivers that a man can carry on his back and use while he walks around. Lynn took Mautz's car and we went out about 10 miles with one of them and they worked fine, and we spent an hour or so just walking around the post in different places talking back and forth as we went.

I was glad to get the brownies and candy – they came through in better shape than they ever have before.

I got a big kick out of some of the enlisted men yesterday. The place where we had our direction finder setup was right close to the enlisted men's school, and I guess some of the men had been asking the instructor in the radio course a lot of questions about what we were doing, so he brought the whole radio class out and had me tell them what we were doing. Those men aren't one bit dumb either; the explanation I gave them was thoroughly technical, and from the questions they asked afterwards I could tell they'd understood it surprisingly well.

4/19/39

Direction finding is coming along pretty well. I was talking to the chief engineer out at the labs this afternoon -- we're working with him more or less on the tests. Some of the tests we have to make call for equipment that I haven't been able to locate so far so I finally decided to build it. I've got to go out to the lab tomorrow afternoon and find out what I can get from them in the way of parts.

4/25/39

I got to figuring out the value of the stuff I'm working with. The direction finder is probably valued at about three thousand dollars, I've got a couple of other sets each worth about a thousand more, and tomorrow I'm getting another set valued at $4800 – that will be about ten thousand dollars worth of radio equipment I'll have signed receipts for, and will therefore be responsible for.

5/3/39

In Jersey they have only two seasons, the hot humid season and the cold damp windy season.

5/17/39 *(the New York World's Fair)*

I guess I told you I was up to the Fair in New York last week-end. I went up Saturday intending to stay until Sunday night. Saturday was wet and foggy and drizzly. I got out to the grounds about four in the afternoon and stayed until almost midnight; then I went back to the Y for the night, and when I got up the next morning it was such a nice day I decided I didn't want to go to the Fair that day – I wanted to come back here and play tennis in the afternoon. So I did. What I saw Saturday was pretty interesting, although a lot of the things aren't open yet. I saw a few television demonstrations, which were mostly pretty satisfactory, and I happened to run into Vox Pop in the RCA exhibit – how long have they been on Saturday nights? GE had an artificial lightning display – ten million volts. They had some pretty interesting fireworks on the lagoon, and a night exhibition of water skiing too.

5/24/39

I think I'll be leaving here about June 12, although if my work isn't quite finished by that time I may stay a few days longer. I don't know how soon I'll be home, though, because I want to stop at Camden and see RCA again, I want to look around New York a little, there are a couple of places in Ohio I'd like to stop, and I may get more ideas too before that time. I might go up into New England too.

5/31/39

Dear folks—

Two more weeks; I just got my orders today saying I can leave here two weeks from today. . .

Just now Miller came in here with a pint of ice cream he couldn't handle by himself, so I helped him. They don't have really good ice cream here in Jersey, though. Nothing like Madison ice cream.

6/17/39 *(from Hotel New Yorker in Manhattan)*

Going to spend the day looking around town here for a job. Saturday I'm going to meet Mautz and we're going up to

Poughkeepsie to see the crew races. Sunday I think I'll spend at the Fair. . .

I didn't enjoy leaving Fort Monmouth nearly as much as I'd thought I would. Too many people I hated to say goodbye to. It's really been pretty nice there in the last year, in spite of all the things I haven't liked. I'd only learned in the last month to like my company commander.

CHAPTER 3

Letters from Flight Schools

August 1939 – May 1940

Following his year at Fort Monmouth, his acceptance for Army flight training, and a visit home, Dad went to Parks Air College, about which close friend and fellow cadet Carl Walter later wrote:

"We all lived in terror of our flight Instructors – the "washout rate" was about 40%. Aldro got airsick on his first flight. He got sick the next day on his second ride. Usually that meant the end – but the instructors apparently saw something in Aldro's grit and ability and gave him a third chance. This time he did fine, didn't have to clean out the plane, went back to his classmates and eventually graduated with them to Randolph Field in Texas."

In November 1939 he was transferred to Randolph Field near San Antonio and in February 1940 to the Air Corps Advanced Flying School at Kelly Field. He graduated May 11, 1940.

Numerous letters make it clear that flying planes back then was a lot more physically demanding then than it is now. Aerial acrobatics were an integral part of training but -- as described in numerous letters -- not necessarily what a mother would be thrilled to read about. And those big white scarves that pilots (and Snoopy) wore back then weren't a fashion statement -- the cockpits were open and it got cold!

Letters from Parks Air College, East St. Louis, Illinois
August – November 1939

8/13/39

Got into Saint Louis last night at 11 PM. I rode from Chicago on the Abraham Lincoln, new streamlined train on the Alton line. . .

The streamliner is a pretty nice train. I rode in a parlor car all the way (courtesy USA), which was all right too. On the way down I went into the club car after a ham sandwich – I'd bought them from porters lots of times – only this one turned out to be a little fancier. First a man pulled up an ashtray by my chair, where I'd told him I wanted to eat it as I read, then he reversed a few gadgets and made a table out of the ashtray, which he spread with an embroidered tablecloth, or printed or something. Then along came a napkin the size of a dishtowel, and a big tray. The tray had on it a glass of ice-water, a slice of tomato, three olives, a pickle, and a triple-weight extra-heavy duty dinner plate on which was one ham sandwich – with the crusts cut off. I was so disgusted I crammed the thing down, paid for it, and went back to the parlor car to read. The train was all air-conditioned, incidentally, and, more unusual, the air-conditioning made it comfortable. . .

Tomorrow, I go over to East Saint Louis to Parks Air College.

8/14/39

Dear folks—

Well, this morning I came out here as per schedule; the assistant field manager came in to St. Louis after me, and the first cadet I saw when we drove up by the hangar was Walter. We didn't get much chance to talk then, because he was due to take off in a few minutes and I had a lot of standing in line to do.

Our room is about 15 by 20 feet (corner room) and there are 5 of us. One of the beds is a double-decker. We have two closets and 2 dressers, with a total of 12 not very big drawers.

8/22/39

Well, we've been flying now since Monday, about a half to three quarters of an hour a day. . . Right now 45 minutes is plenty for most of us, because we get tired easily. Later we'll get more, when we get more used to it.

We always fly dual now, of course – that is, with an instructor. I think we'll solo in a couple of weeks. . .

We have ground school in the afternoons. We have a course in federal aids to navigation (beacons, radio beacons, etc), a course in theory of flight, and we'll start meteorology, maps, and engine courses pretty soon. Right now they're running a refresher course in mathematics, which most of us don't have to take.

I'd like to have you send me my old blue sweater. Ordinarily it's not cold flying, but some mornings it's chilly even on the ground and when we get up in the air, maybe 5000 feet or so, and stay up there almost an hour with a breeze flapping around, a sweater would be comfortable. We always wear scarves. We have big white silk ones that we can get all bundled up in.

9/4/39 *(Hitler and Mussolini)*

Tomorrow and this week we fly mornings. The air is smooth then. We're practicing take-offs and landings now. . .

The European business seems kind of disgusting right now. I haven't read about it anywhere, but I heard somewhere that Mussolini is trying to get Hitler to call the whole thing off. Mussolini's been doing the most surprising things lately.

9/6/39

We're flying about 1½ hour a day now. It's easy to get tired; ¾ hour in a session is enough. We go up twice.

9/13/39

Dear folks—

Well, flying today was a real workout. The air was the roughest and the wind the shiftiest it's been since we got here. We'd fly along level at 400 feet and suddenly drop 50 feet or go up 150 feet for no apparent reason at all. Vertical air currents do that and there was one stretch of air where flying was just like riding horseback. I'm glad my stomach stopped bothering me. If it hadn't, I'd have been a wreck after today's flight. None of the fellows get sick now, although some of them took a couple of weeks to get over it. . .

Uniform here is pretty informal. We wear coveralls. Polish on shoes is optional – in fact, shoes are optional. About half the class goes to ground school and mess and all over in bedroom slippers, and take them off while they sit around.

Some of the fellows here are characters. A fellow from the Tennessee mountains said "You folks up north wear shoes a lot, don't you?" After the uproar had died down, it developed that "shoes" meant conventional shoes and slippers, sandals, sport shoes and the less dressy types go by other names. Another fellow – two of them in fact – are from the Ozarks, famous for hill-billies.

9/20/39

The last couple of days we have been getting started on some of the simpler aerial maneuvers -- sharp turns and chandelles. Today the instructor looped once but he didn't have me do it. Yesterday he had me practicing those maneuvers and every once in a while he'd cut off the motor while I was in the middle of something -- maybe pointed straight up or lying on my side or something -- the idea being for me to practice getting out of situations like that without losing control.

We do a lot of that. Very often when we're practicing something -- maybe at high altitude, more often at low altitude -- the instructor will cut off the motor and then we have to pick out a field to land in and head for it just as we'd have to do if the motor should really quit and we'd have to make a forced landing. In these practice forced landings we never actually land. About 20 feet off the ground the instructor opens up the throttle and we go back up again. We just get down close enough so we know we could have made it if we had to.

9/27/39
Dear folks—

Feeling kind of exuberant tonight – I just had my second check flight this afternoon, and passed it OK I guess. It was a pretty interesting ride; the lieutenant, on the way back to the field, looped the ship once. We've never done that yet ourselves. . .

Take it easy – and watch that jaywalking. Really, I think you're in more danger when you cross streets than I am when I'm flying.

73

Aldro

9/29/39

We had another washout yesterday; that leaves 24 of us, out of 41 originally. No – I think it's worse than that – I've lost count. I think we lost two yesterday.

We've just been having a bull session with our two Ozark hill-billies. Now there seems to be an argument in one of the other rooms about who won the Civil War. I didn't know we had any rebels here – must be some fellows in the new class.

10/2/39

My room-mates are Toner and Hanlon, both from Massachusetts. Hanlon went to Harvard, as is easily noticeable in his speech, and neither of them has any kind of a mechanical, practical, or even tinkering background for the kind of work we do here, and consequently I do as much instructing on the side as the instructors do in class.

I guess we'll be here six weeks more and then go to Texas -- those who make it. I'll see Walter there, of course, provided we both hang on. . . I have an idea we'll get out of here just about the right time, too, if present weather is any indication. I haven't heard from Walter yet so I don't know how things are down there. I know the kind of airplanes they'll be flying, though, and they really ought to be fun. They're different from ours – faster, and they have radio equipment. They have closed cockpits too, instead of open ones like ours. . .

10/12/39

I should be feeling pretty good tonight – and I guess I am – except that I'm awfully tired too. Yesterday I passed my third check; today my instructor told me I got a "C" on it, which is the best grade they give around here on flying. If you get a D you're all right; if you get an E you get another chance, and if you get an F you go home. A C is very unusual and to be regarded with due respect and awe. Well, anyway, after getting a C on that check yesterday, today I got another one on spot landings. Spot landings aren't easy; you might think that all you've got to do is fly down to the spot where you want to land and put the wheels down on the ground and then coast to a stop – something like stopping a car on a particular spot. But that's not it at all. What it could be compared to is getting going in a car, then deciding on a spot where you want to stop (and it can't be straight ahead) then disengaging the clutch and coasting to a stop on that spot without using engine or brakes. Landing an airplane isn't just flying down to the ground. You have to close the throttle so the engine is just idling and isn't giving you any power; then you glide around and arrive over the spot where you want to land at an altitude of about 2 feet and barely flying speed, and then you coast along in the air till you lose flying speed and drop to the ground on the wheels and tail wheel. Well, maybe I shouldn't have gone into it.

10/18/39

It's been pretty chilly here the last few mornings but I think it's warming up this morning. I'd been up at 3000 feet for most of my first flight and then just before I came in I went down to 400 feet to practice a couple of low altitude maneuvers and nearly froze. In the space between 600 and 500 feet I think the temperature dropped 20 degrees. That, incidentally, is not a very unusual condition in the morning, except it's not usually so noticeable. We get pretty well warmed up anyway pushing and wrestling a 2200-pound airplane around the sky doing acrobatics at speeds from 90 to 160 miles an hour. I was really worn out after an hour's flight this morning. . .

Windy days aren't bad for flying, unless it's so windy that planes on the ground blow away. In the air it doesn't make any difference at all and it makes landing and taking off easier because it has the effect of lengthening the field and giving more room. . .

Three fellows washed out today on checks; that leaves 18 of us, out of 41 originally.

We'll have I think one more check to pass. . .We'll have to be perfect on all the maneuvers we've had so far and we'll have to do fairly well on our acrobatics, too, which we just started. The catch is that acrobatics, I've heard, very often roughen up the rest of a fellow's flying, so the fact that we may have done all right to date doesn't mean we'll still be able to do the old stuff as smoothly as we used to.

10/22/39

Yesterday I saw "Mr. Smith goes to Washington;" it's awfully good. . . After riding around in cars and street cars, I think I'd rather fly in the roughest air we've had than ride in a car on a smooth road. Bumps in the air are gentle. They don't jounce you around. You just gain or lose altitude unexpectedly, or maybe get your wings rocked around a little, but it's less troublesome than the slightest jounce in a car.

10/25/39

We've started having a little fun in ground school. We're doing troubleshooting now in our engines course instead of just sitting around listening to someone talk about magneto theory who didn't know anything about magneto theory. Troubleshooting's fun; the instructor puts some kind of trouble on the engine, then we start it up, listen to it, and try to find what's wrong. It's the first chance we've had to get our hands dirty since we got here. We've got a couple of 600 horsepower engines we use for that.

Acrobatics aren't as bad as they sound; they're fun when you get so you can do them fairly well. I didn't like it the first few times I was on my back, but that was only because it was hard to figure out what was going on. It's easy to get tired doing acrobatics, though, because it takes heavier control pressure than in ordinary flying.

Randolph Field, where we'll go when we finish here, is near San Antonio. That's about the same latitude as New Orleans.

10/31/39

Folks, tonight I feel like I did the day I first soloed. I'm going to Texas. I had my final check yesterday and today my instructor told me I'd gotten a C on it and that the officer who checked me had told him I'd given him the best acrobatic ride he'd had since he's been here. . . This morning someone flew in from Randolph Field in a BT-9, a ship like we'll be flying when we get there. It's a low-wing monoplane with 400 horsepower engine, radio, and all the gadgets. You should have seen the crowd around that ship as it was standing out in front of the hanger -- fellows who'd passed their final checks all standing around imagining already what the controls felt like in her.

Sunday

Tomorrow, we leave, I guess. . .

We finished up our 65 hours last week; I finished mine Friday. Thursday, I rode in the front cockpit, my instructor in back, the opposite of the way we usually ride. Everything looks and feels different that way. We did a few acrobatics I hadn't done before, and played around in general; lots of fun.

Letters from Randolph Field, Texas
November 1939 – February 1940

11/11/39

We got here Wednesday afternoon, but I didn't write before because I didn't have anything to write with. . .

Truax had a clipping from the Journal and I saw our pictures. The glossy print from the photographer was bad enough; I didn't think the engraver could possibly make it worse, but I guess he did. And I was smiling – at least I thought I was, and I should have been, because the picture was taken just after I'd passed my final check and been complimented on my acrobatics.

I guess we won't do acrobatics here; emphasis will be more on precision cross-country flying and navigation and things. You ought to see these airplanes, though. They're really something.

11/19/39 *("We" was written by Charles Lindbergh)*

Yesterday the post commander talked to us, and we had a safety lecture -- -- not about airplanes, but about cars, because the air corps loses more pilots in car accidents than it does in airplane accidents. About twice as many. . .

The place is full of Mexicans, it seems. You see little kids – girls too – running around the downtown districts in knee-length leather boots. Not lace boots, but more like riding boots. No socks – just a dress, bare legs, and high boots. Looks awfully funny. Occasionally you see a dark-skinned Mexican woman with her face all powdered up and she looks like a corpse.

So, you read "We." I happened to run across it in a secondhand bookstore in St. Louis where I was looking for something else. The first solo flight is definitely an experience. I had mine after about 10 1/4 hours, which is average for a course like ours. Lots of civilian schools, training in "flivver" planes, will solo a student after as little as 2 or 3 hours of dual instruction. Army regulations don't allow solo under 8 hours. That's a feeling, though. As soon as my instructor got out of the cockpit and told me to "take her around" the whole airplane seemed to feel different. It even taxied differently, on the ground. It felt lighter, and more responsive to the slightest control pressure. After I gave her the throttle and started to take off that feeling disappeared entirely; I circled the field, landed, bounced a couple of times, then took off and circled the field again. I think instructors get as much kick out of their students' first solo rides as the students do; anyway, mine was grinning all over his face when I taxied back up to the hangar. Mine had been the second solo in a class of 40 students. That doesn't mean much, though.

Incidentally, I found when I got here that I had reputation here ahead of me. Somehow, the story of the check pilot's comments on my acrobatics on my last check flight at Parks had gone on ahead. Now I've got something to live up to.

11/24/39

These ships are really something. The cockpit is entirely closed in and that, in combination with a fairly warm climate, makes warm flying clothes unnecessary. In fact, it can easily get too warm in the cockpit, but to take care of that we have a little adjustable ventilator that really works, unlike most of those in cars, busses and trains. As for gadgets -- the pilot has 23 instruments and gauges to watch and about 25 different gadgets to work -- no, there are more than that. Anyway, it's enough to make things interesting.

We started flying Tuesday. Didn't have much fun that day -- flew around for about an hour just looking at the country around the field. The ships were entirely new to us, of course, and we had absolutely no feel of the controls. Flying in a ship that you can't feel isn't much fun. The next day was lots better, though, and we'll probably start soloing next week. It may sound funny, but the best single way to get the "feel" of an airplane's controls is to take it up to a safe altitude and then stall it and see how the controls feel when they're not having any effect. That's what we did Wednesday. We flew around 6000 feet most of the time. Altitude's pretty hard to judge right now; 6000 feet here felt and looked like about 2000 feet over East Saint Louis. The country's very different, that's why. . .

This place is full of Southern accents, all different. And what isn't Southern is New England -- Harvard, etc. Doesn't seem as if there's an ordinary American in the place. You ought to hear Boston and Alabama fighting the Civil War -- Boston moaning about all the "nigger-talking" Southerners and Alabama coming back with remarks about the "nigger lovin' damn Yankees". . .

You shouldn't worry about me having to get used to a new field. There's no comparison to Parks. Parks airport is about one tenth of the area of our field here. Anyone that can land an airplane at Parks couldn't possibly miss this field. It's easier to see from a distance, too; Wednesday we got out about 15 miles away and we could easily see the field; at Parks, five minutes after taking off we'd be out of sight. The air was always smoky around there, too; sometimes it was so thick that in places we couldn't see the ground from 400 feet. I've landed there when I couldn't see the field until I was down to 200 feet. It never gets that way here. Maybe I shouldn't have told you about that. It's not as bad as it sounds. I always knew where I was, because even though I couldn't see the airport or the ground directly below me, there were other things I could see that told me where I was.

12/1/39

Wednesday, I soloed, after 4 hours dual so far. These ships land much more easily than those we flew at Parks. Back there it took me nearly 20 hours before I could consistently land without bouncing; in these ships, with 4 hours I've bounced only once. They roll smoother too, along the ground. It's a little hard to catch on to the increased speed, though; you don't have much of a sensation of speed and you just have to know when to start a turn to come out where you want to when you want to. . .

Had a letter from Wayne yesterday; he's working on a new photographic Christmas card which sounds pretty interesting. Makes me wish I had a dark room here. Things get awfully tiresome at times with none of that stuff around. It's not much fun to take pictures when someone else is going to develop and print them. It's more expensive too.

12/5/39 *(Cicero, Ciceronis, Ciceroni is an example of Latin declensions)*

I'm always under more tension when I'm in a car – driving or not – than I am in an airplane. I feel much safer in an airplane up away from the crowds where I've got only my own mistakes to contend with and don't have to worry about what every other halfwit in sight is going to do. . .

I lost several days flying because one day I went over to the hospital to get something for a sore throat and they, being low on business, put me to bed. I wasn't sick, but they wouldn't let me go until yesterday morning. One fellow there had taken an eye exam, flunked it, and while they were waiting to give him another they put him to bed too. He went there last Thursday and I guess is still there. They like to have cadets over there because they make about 200% profit on their food. . .

We're starting code practice; this morning I took a test to get out of it. They require only 12 words a minute and I can take 25. Instead of getting out of the course, they told me they may make me teach it. They're short of instructors.

Cicero, Ciceronis, Ciceroni – how time flies!

12/11/39

I think I told you once there were a lot of Mexicans in San Antonio. That was an understatement. Saturday night I missed a bus, so while I waited for another I took a walk down in a section of town, not far from the main business section, where I hadn't been before. In an hour I didn't hear an English word spoken. Stores opened right on the streets – it would be hard to say whether they had doors or not – and all along both sides of the street the sidewalks were sheltered completely by porch roofs, kind of, sticking out over them. Signs were all over the place, all Spanish but not hard to guess the meaning of. One product was advertised "Satisfaccion absoluta.". . . I've told you about little Mexican girls coming to town all dressed up in their Sunday boots. I saw boys, too, four and five years old, possibly, wearing ten-gallon hats – not the drug-store variety, either. One little kid who couldn't have been more than four years old, wearing a big buff-colored hat, was sitting in a shoe-shine place getting his cowboy boots shined by a kid three times his size. I saw a man, incidentally, the only one I've seen so far dressed in traditional though serviceable cowboy style: cowboy boots, pants, wide metal-studded belt, black embroidered shirt, bright-colored silk rag around his neck, tan face, needing a shave, ten-gallon hat. No chaps or six-gun though! Also no spurs. Farther down the street I think must have been a Mexican celebration of some kind. Cars were parked all around a big building, lots of people inside, and outside, standing around alone, in groups of two or three, or with other people, were men dressed in ceremonial Mexican style, with big colored silk scarves and most of them carrying big guitars, some playing to themselves, quietly, as they stood around or waited for traffic to move. Firecrackers were going off in the park across the street. I've got to go back there with my camera some day. . . There's a small stream through the middle of town, too, with banana and other palm trees along the banks, that I want to get pictures of.

One of my roommates is always getting letters from, or talking about, kids he knew in reform school. I get a lot of fun listening to him. He used to work in a reform school, as athletic director or something. He's a psychology graduate from Yale, and played football there.

12/19/39

Dear folks—

Well, I'm sure now I can't get home for Christmas. We get four days off; that's not enough. . .

I've just come back from night flying. It was my second night; we get three. It was moonlight, quiet, all the stars were out, and the air was perfectly smooth. I flew at 2500 feet part of the time, and partly at 1500, occasionally dropping down and making a landing with wing lights. Everyone always gets more fun out of night flying than any other kind, especially on nights like tonight. Everything on the ground showed up in the moonlight; I could see cars driving up and parking in front of a hotdog stand that I was circling near most of the time, lights -- dim lights, probably kerosene lamps -- in farmhouse windows. Mesquite showed up as shadowy splotches on the ground. San Antonio was a big area of bright lights; below me someone was loafing along the highway in a car on the wrong side of the road, and I was glad I was up where I didn't have to worry about guys like him. Off to both sides I could see the navigation lights of the other ships, circling around in their assigned zones, and on the radio, I heard, riding over a Mexican musical broadcast in the background, the flight commander telling them, at the right time, to come in and make a landing.

12/23/39

Here's a picture of the heavy flying suits we wore the last few bone-chilling days at Parks. We don't need anything like that here, because in the first place it doesn't get so cold and in the second place these airplanes are warmer. Those suits were all leather, fur lined all over, and they had big heavy fur boots which the picture doesn't show.

Wayne was really prompt about my photo stuff. Matt's going to phone me when it comes and we'll set it up in his bathroom. He's got a bathroom big enough to sleep in. I've found it's not only no fun and too expensive having my pictures developed commercially, but besides that I can't get a decent job done. They underdevelop my films and don't focus their enlargers right and then use stale developer on prints. It's disgusting. There's only one way, it seems, to get decent work done on miniature camera films and that's to do it yourself.

We've been pretty busy flying this week; we're the only class here right now and we've got all the airplanes we can use. Thursday I flew nearly four hours. We start flying again Wednesday. I had my first instrument or "blind" flight Thursday, too. Lots of times the airliners have to fly when they can't see the ground, you know; that's called instrument flying, because you use your instruments to tell you where you're headed. My instructor rode in front and I rode behind with a hood over me so all I could see was the instrument panel; nothing outside. Then he'd tell me where he wanted to head for and at what altitude, and my job was to do it. It's very different from ordinary flying.

I hope you have a nice Christmas.

73

Aldro

12/29/39 *(Sonja Henie has won more Olympic and world figure skating titles than any other ladies figure skater)*

Dear folks—

I should have written you sooner after opening my Christmas presents. Thanks a lot for the socks and handkerchiefs and things – and I was just about out of pajamas. The brownies came through in good shape a couple of days later, but the lefse was all moldy. I guess it's not a good idea to try to send stuff like that so far. . .

I remember last Christmas eve I wanted to go to bed early but instead I let Margaret coax me to go to some kind of a show with Sonja Henie in it, and went to sleep – nearly – in the theater. . .

Aunt Jo sent me a tie, Drevdahl's a dollar. . .

One of the other flights is flying tonight – someone just flew over with his engine wide open turning up about 2300 revolutions per mintue. Airplanes always take off wide open, but normally you throttle back as soon as you get about 50 feet up. Every once in a while someone forgets to throttle back and goes merrily on his way at full throttle. When a propellor turns so fast that the tips are moving through the air at about the speed of sound, the noise increases tremendously, but the pilot, for some reason, doesn't usually notice it. But the whole countryside for miles around can hear it, and it's a really penetrating sound. I wouldn't like to be a baby here.

I think we start acrobatics this week. We did a few snap rolls today. It's been chilly enough so we find it best to fly with canopies closed. . .

Happy new year -

73

Aldro

Jan 4, 1940 *(mis-dated 1/4/39)*

"My instructor started me doing slow rolls a couple of days ago and I'm all sore from it. It takes all the strength I've got to get around, and then some. Have to use both hands on the stick and heave as hard as I can. Today I had a ship that rolled a little easier and I didn't get sore from it."

1/10/40 *(mis-dated 1/10/39)*

We've been having a lot of instrument flying lately. The student rides in one cockpit, covered by a hood so he can't see anything outside the cockpit, and the instructor rides in the other cockpit to see that the student doesn't get into trouble on account of being blind. Then, using the ship's instruments to tell him what's going on, the student does turns, climbs, glides, straight and level flying and things. It's very tricky. Not being able to see, you get all kinds of funny feelings about what the ship's doing, and you just have to learn to believe the instruments, regardless of what you think the airplane is doing.

I heard a story today about some guy out on a cross-country flight, flying on instruments between cloud layers. He could see around him for some distance, but he couldn't see sky, ground, or horizon, so he of course had no way of knowing how he was pointing except by instruments. Well, he happened to see someone ahead of him, going in the same direction but slower, and also on instruments, of course. So he rolled over on his back and flew upside down past the slower ship, and the pilot of that ship wondered what was this anyway and decided he himself must be upside down -- instruments or no instruments -- so he rolled over on his back too!

Most airplanes won't fly on their backs more than a few seconds before the engine quits. The gasoline feed system doesn't work that way. Ours won't, and those on the PT13's we flew at Parks wouldn't either. Engines would quit in about 5 seconds, and then we'd have to get right side up again before she'd start. . .

Radio in airplanes is definitely handy. Yesterday, just as I was starting to taxi out from the line to take off, a mechanic came running up to my ship, climbed up on the wing, reached in the cockpit and grabbed my microphone. He'd happened to see another airplane out on the field, going to take off with his baggage compartment door open. So, he gave the man a call on my radio and stopped him. It wouldn't have been dangerous, but the door would probably have been torn up by the breeze.

Jan 14, 1940

One day last week my instructor was feeling frisky and showed me a lot of fancy acrobatics – things we're not allowed to do solo, but just for interest. The only acrobatics they let us do solo in these ships are slow rolls and snap rolls. Loops tend to be mushy, and we don't do them. Well, anyway, he was showing me double snap rolls mixed in with other maneuvers – it got kind of complicated. Then I did snap rolls till he got sick. It's the first time I've ever flown an instructor sick without getting it myself. He's always showing me snap rolls, and he can't do them very well. I think he shows them to me so he can get the practice. They can be awfully aggravating at times – you see, the airplane is actually out of control in a snap roll, and the quality of the performance depends on how accurately you time your regaining control.

1/22/40

The airway weather services are really good. If you're taking off for someplace and expect to get there in an hour, or maybe 3 hours and 55 minutes or so, they can tell you nearly exactly what weather you'll have when you want to land. They're not like newspaper weather reports. The airway weather service has more complete data, better trained personnel, and can tell you nearly 100% what kind of weather you'll have all along your trip.

My instructor was away last week; he'd gone to Langley Field in Virginia to bring an airplane back here, and he was held up by bad weather in Louisiana. So – I flew by myself for a week. Passed a check ride and a spot landing stage, and had an instrument ride with someone else's instructor; except for that, I was an orphan.

Feb 1, 1940 *(parachutes aren't just for jumping out of planes)*

Your fears about sneezing in an airplane are kind of exaggerated. I read that too, somewhere, but actually it's not as dangerous as sneezing while driving a car, because although you go a lot farther during a sneeze in an airplane, there's not very much you could run into in that time. The sky's pretty big.

Weather's been warm here last few days, since about last Saturday. I guess yesterday it was in the 50s or so, and when this fog lifts, it will be higher. Speaking of weather, we've been taking a meteorology course. For the last few weeks and a few days ago, a list of names was read off in class with the statement that we'd made the highest grades in the course (whoopee!). Then came the statement that in recognition of our efforts, we would have the distinction of presenting before the class short talks on any subject we liked which was related to the course. That's the army. As a reward for hard work (well -- mostly luck) we get more work. . .

I've got a parachute I'm going to send you, from a parachute flare like we use for night landings sometimes. A flare is a kind of glorified candle dropped from the airplane and automatically lit, and hung on a small parachute so it will stay up long enough to give us time to get down first. Then after the flare has burned out, the army doesn't bother to recover the parachute. They've got a lot of silk in them, and they're pretty popular around here for making dresses and things. . .

I've taken several pictures from the air the last few days. Yesterday there were some very pretty clouds in the sky -- they seemed about 15,000 feet up – so, having a 3-hour period to fill and not an awful lot to do, I climbed up after them to try to take a picture of Randolph Field through them from above. I didn't, though. I climbed to 15,000 feet and didn't reach them, so I gave up. I wasn't entirely disappointed, though, because someone had told me a BT-9 wouldn't go above 10,000 feet, and now I know better. I think she'd have gone to 20,000, if I'd insisted, but I didn't because above 15,000 feet you have to have oxygen equipment and I didn't want to risk a temporary faint.

Those prize county plays sound interesting. I remember seeing some of those in Mount Horeb once, and I seem to remember a play you were in over at the schoolhouse when we lived on the farm. Didn't we fix you up some whiskers out of some car cushion stuffing? I don't remember how we fastened them on. I do remember there was a washing machine in the play. That seems an awfully long time ago; longer than anything else I can remember.

Latest news around here is a measles quarantine. All men in each company who haven't had measles or are believed susceptible have been moved to a separate part of the barracks, assigned to separate tables in the mess hall, ordered to stay away from the theater, post exchange, and like places awaiting developments. I think there are about 40 cases on the post now; no cadets that I know of. It's quite a joke among the isolated ones; they call themselves the "leprosy squad." They're not going to like it one bit, though, if, as seems likely, they're not allowed to go to the big dance in town this weekend celebrating completion of our course here and transfer to Kelly Field.

I almost forgot that I think we move in about a week -- maybe 10 days. Kelly Field is just on the other side of San Antonio; mail sent to me here will be forwarded. Take care of your colds.

73

Aldro

Feb 7, 1940

Dear folks—

Just two hours and 40 minutes more flying time here; that will be tomorrow and the next day. Monday, we go to Kelly Field. Kelly is over on the other side of San Antonio, just outside of town. It looks quite a bit closer to town than Randolph from the air, although San Antonio stretches out pretty far in that direction. . .

Last Sunday afternoon five of us went into town and went horseback riding; for three of us, it was the first time. It was fun, but oh. We were out only half an hour and I'm not over it yet. It hurts to sit down and I'm too stiff to get up. Even lying down brings out a sore spot somewhere. The other fellows are as bad. I think riding shouldn't be as hard to learn, except the toughing process. It's very cheap here -- $0.50 an hour. I don't know what it is in Madison, but the boys from New England say it runs about $1.50. . .

73

Aldro

Letters from Air Corps Advanced Flying School, Kelly Field February – May 1940

2/14/40

These ships aren't awfully different from the BT-9… the most outstanding difference is that these ships (BC-1) have retractable landing gear. In flight, the wheels can be pulled up into the wings so they're not sticking out in the breeze holding you back.

2/19/40

We only flew one day last week . . . This airplane (BC-1) is a lot like the BT-9 – I think I told you all that. It has 550 horsepower instead of 400; 550 horses lifts you right off the ground in a hurry. Besides not taking so long to get off the ground, the BC-1 doesn't need so much room to land either, which is kind of surprising because its landing speed is about 20 MPH higher than the BT-9. We land these ships about 90 miles an hour. One thing that's really nice – the BT-9 and PT-13 both had very grabby brakes; sometimes if you just touched them they'd take hold and nearly throw you over on your back if you were going very fast. Well, no one ever gets hurt that way but it's very humiliating. But the brakes on the BC-1 don't act that way. There're smooth-acting, and don't grab . . .

I saw "Gone with the Wind" yesterday. With an intermission, 4 hours wasn't bad but it was enough. I liked it though.

Feb 28, 1940 (*written on stationary with motto: "Ut viri volent")*

My Latin's a big help here. I can translate the motto at the top of this stationery. Caesar comes in handy too, because life as a flying cadet, like old Gaul, est omnia divisa in tres partes -- on the ground, in the air, and off the post."

(The first sentence of Caesar's Gallic War, a standard Latin text, is: "All Gaul is divided into 3 parts.")

Mar 7, 1940 *("Ut viri volent" means "That men may fly.")*

The family's Latin is deteriorating; Elizabeth, look in your vocabulary again and I think you'll find "volent" to be a subjunctive form of a verb meaning "to fly." And it's nearly six years since I took Latin – so there! Teacup brain!

3/14/40

I'm not in the regular army; when I finish here my reserve commission in the Signal Corps will be transferred to the Air Corps and I'll be put on active duty for a while. I'll take about a year of it, I think.

We just got some new microphones for the airplane transmitter. They're throat mikes; you wear them around your neck and they pick up the sound directly through your throat, by contact. They're awfully handy, because they leave your hands free for flying.

3/20/40

Dear folks—

We haven't been very busy this week . . . The other class graduates today, and that's why we didn't fly yesterday; they were using all the airplanes to practice the aerial parade they put on for their graduation today. That was something; 120 airplanes all flew across the field in formation. It was good formation flying too; the best I've seen here.

4/10/40 *(He paid $35 for his first car)*

Dear folks—

Just got back from a time and distance problem this afternoon. An instructor started out from a town about 60 miles south of here, headed for Kelly at an announced time; and several of us were sent out to different towns, from there to come along and pick up the instructor at assigned points along his route. I went out to Yoakum, about 70 miles east of here, then came back to Floresville, about 30 miles southeast to meet the formation, then flew back here in formation.

We were supposed to go up to Dallas this afternoon and come back tonight, but the ceiling's too low. They don't like to send us on cross country when the ceiling's less than 5000 feet; you can't see very far from low altitudes and navigation isn't so easy. Night cross country flying's lots of fun. I had a 200-mile triangular flight Monday night -- in a brand new airplane too. They're just getting in some AT-6's; they're a new ship nearly the same as the BC-1, except they're all metal, have the gadgets arranged a little differently, land slower, and fly a little faster. I'd had my first ride in one that afternoon; 20 minutes, dual for familiarization. Then I took her off that night solo on cross-country. It's an awfully nice ship – smooth, and with the canopy closed, the 600 horses up front sound like a couple of kittens. Night cross-country's beautiful flying. The air is perfectly smooth and you can see beacons for 40 or 50 miles on ahead from comparatively low altitudes. . .

Don't you know what makes an airplane go up? The propeller pulls it forward; you know that. And as it moves forward through the air at high speed, the wings are built so that the air strikes them harder on the under side than on top and that holds them up. You must know that too. . .

Did I tell you I bought a car? It's a 1936 Oldsmobile sedan. I'd been looking around for quite a while without seeing anything I cared for at all, but I think I've got something now. From what I can tell so far, she's in good shape throughout, runs about as cheaply as anything, and handles very well. . .

Three of us drove out Sunday to a little town in the hills about 20 miles west of here. Took a few pictures, nearly went swimming only we didn't know there was a place so hadn't brought our suits, cruised around the hills awhile, saw the eclipse, looked at an old Spanish cemetery with graves dating back to 1820, then came home. .

Well – enough rambling. Take it easy.

73

Aldro

5/9/40

Had a cross-country up to Abilene yesterday afternoon; it's pretty country. I've never seen such a variety before on any trip. Right around here is ordinary farm land; about 15 miles north it gets hilly and quite rough, with mesquite growing along hillsides in contour lines, making the ground below look like a contour map from the air. Farther north it flattens out a little into just ordinary level desert, with mesquite scattered around and only an occasional stream bed, dried up and glaring white in the sun. . . Then farther north still, by Abilene, the ground is flat all over, not quite so dry, and scattered over the landscape are isolated cliffs, rising steeply from the level ground to 100 or 300 feet and sharply level on top. I've never seen anything like that. I think they're called mesas.

5/13/40 *(This follows graduation, which Margaret attended)*
Dear folks—

Well, I guess it's about time I was writing. The girls left at 12:30 with Owens -- 2 hours ago. Since then I've been out to Kelly, had a group picture taken of the 17 of us who came to Randolph from Parks. The picture is as nearly identical as we could make it to the one we had taken when we left Parks; only now we're in officers' uniforms instead of coveralls. Now for 10 days I haven't a thing to do; just stick around.

They got here earlier than I expected; about 10 Friday morning. I couldn't get away until 3:30, but then we went out to Randolph Field, looked at BT-9's, looked over the post, then went out to New Braunfels. At Randolph the girls were into everything; they had their pictures taken everywhere, in and out of airplanes and even on the driver's seat of a tug -- a little tractor used for wheeling airplanes around in and out of the hangars. . .

Saturday morning the girls were out to graduation and in the afternoon, after I was through getting paid, moved out, and things, I picked them up and took them horseback riding at Kelly. The horses they had were slow and sleepy, and mine wasn't much better; anyway, they got along OK. Afterwards we came on into town and looked at things in souvenir stores. You should see the Mexican straw hat Margaret bought. We went up to the top of the tallest building in town, too, and looked at the view. It was pretty.

Well, I've never given you the idea I could take them riding in an Army airplane -- and I can't. At least not here, now; it is allowed, within very set limits, but they wouldn't let any of us do it now because there were too many people per pilot and too few airplanes per pilot. Anyway, Sunday morning we went out to Stinson Field (Municipal Airport) and took a 10-minute ride over the city in a cabin ship. I couldn't have taken them myself in an airplane of a type I'd never flown before, even if I had been able to get around civilian license regulations. . .

On the way back from the airport, we looked over one of the old Spanish missions and took a few pictures. Then we went out to Bandera, about 50 miles northwest, to a dude ranch. . . and went horseback riding. That was fun. We had Western saddles, for one thing, the most comfortable type made, instead of the cavalry saddles they had at Kelly. The girls had horses that were gentle, but not dead; they'd trot, if you wanted them to. The horse I got was kind of frisky. He was alright headed away from home, but once we headed back, he seemed to smell the old barn door, or something, and had only two gaits: a trot and a canter -- nearly a gallop. I couldn't make him walk. In order for the girls to keep up with me, I had to let the horse go a few 100 yards, then either stop and wait or turn around and go back to meet them. It was hard making him stand still, too; if I tried to hold him with the reins, he'd rear up. Anyway, we had fun.

5/18/40

"Today I found out I'm going to Langley Field, Virginia."

CHAPTER 4
Letters from Langley Field, VA
May – November, 1940

After graduating from the Air Corps Advanced Flying School, Dad drove home for a visit, and on May 28, 1940 he reported to Langley Field near Washington D.C. Letters from there provided more commentary – often amusing – on flying. He was sent to Puerto Rico in November.

5/28/40

I got to Langley about 2:30 this afternoon in good shape, and am getting settled.

The Blue Ridge Mountains are awfully pretty. I took a few pictures, but they probably won't show up the way they should. Mountains are hard to take pictures of. . .

Oh-oh – here comes a mosquito! Wonder what kind of a time we're going to have with them. We're practically on the ocean.

6/3/40

"I'm assigned to the 10th Bombardment Squadron of the 25th Bombardment Group. We're a heavy bombardment outfit, but because the Army never did get around to getting very many heavy bombers, we have only half as many heavy bombers as we're supposed to have (4-engine 19-ton ships) and make up the difference in medium bombers (2-engines). . . We're to learn to be co-pilots in both types. . .

Someone's putting a P-40 – one of those new pursuit-planes with 1200 horsepower that will do 400 miles an hour --- through its paces outside… From the sound of that engine he must be having a grand old time … there's probably no thrill like sitting behind one of those 1200 horsepower engines with wings stuck on it and almost nothing else – they call them a "flying motor" – and tearing up the sky at 400 miles an hour."

6/9/40

We've been getting settled a little more in the last week and started flying Thursday. . . I'm copilot on a B-17B. That's a four-engine ship like at the top of this page. It's called a "Flying Fortress"; actually, it's like a "flying hotel" inside. You should see the pilot's compartment. The seats are like nothing else -- they're adjustable up and down, which is standard, but also forward and back – sliding, I mean, not tipping -- besides the rudder pedal being adjustable to fit the length of the legs of the person using them. The pilot's compartment is all upholstered -- the whole ship is, I think -- and has rugs on the floor. Pilot and copilot can call any of the crew by telephone – navigator, downstairs in the nose compartment, radio operator back in the middle of the ship, or mechanic, who usually rides way back, and they can talk to each other by telephone too, although the ship is well enough soundproofed so that they can talk back and forth comfortably by just raising their voices little bit. . .

Thursday, we had our first ride. We're short on big ships, so we have to take turns co-piloting while the regular pilot stays in charge. After I'd finished my turn at the controls I went back in the rear cabin and there was the regular co-pilot stretched out on a bunk sound asleep; the chief mechanic was sleeping on the bunk across from him, and farther forward the radio operator was sleeping, too. So, I guessed I hadn't been as rough on the controls as I'd expected to be my first time, since they were sleeping OK. Or maybe I rocked them to sleep. . .

We flew around Chesapeake Bay for a while and over to Richmond and back. While we were over Richmond we happened to run across a couple of Navy pursuit ships up flying around not doing much of anything; they started playing around and diving at us, so the captain just gave the throttles a shove and we ran away from them.

This airplane's a whole lot faster than it looks. It doesn't cruise very fast normally, but it's got plenty of speed in reserve. . .

Maybe I ought to tell you what a co-pilot does. His main job isn't flying, although he has a set of controls and takes over whenever the pilot gets tired or wants to do something else. The co-pilot's main job is to operate the engines, keep them synchronized, keep an eye on all the instruments, and keep track of the gadgets so the pilot can give all his attention to flying the airplane. . .

Friday, we flew down over Dismal Swamp at 4000 feet and then back up the shore. . . Coming up the shore we could see sunken shipwrecks on the bottom just off shore at several places. . .

Kitty Hawk, where the Wright Brothers made their first flight, is down the shore just a few miles beyond the farthest south we've been. I'd like to see that place from the co-pilot's seat of the B-17B. .

My quarters are two rooms with a bath between. They're built for two, but both the rooms are smaller than most of the rooms in the building, so I think I'll have them both to myself. I'll use the other room for a radio room.

June 12, 1940

"Coming down from 12,000 feet I flew through an upper rain shower. Did you know it sometimes rains up above, but the rain evaporates before it reaches the ground? I've seen it before, but last night it surprised me.

6/21/40

I just found out this morning that I've been in the hospital all week. Monday morning I stopped in there to get some stuff to stop a case of athlete's foot that was trying to get started on me, and apparently someone made a mistake and entered me in the records as being sick in the hospital, and no one noticed it until this morning one of the sergeants who goes along on most of our flights happened to see it, and was able to convince the clerk in the orderly room that I couldn't very well be in the hospital because he personally had been flying with me every day this week. Just one of those things. Reminds me of the time at Kelly when they grounded me one day because another cadet with a name something like mine had an abscessed tooth.

7/17/40 *(precision dead reckoning navigation)*

This morning I acted as navigator on a 200-mile flight down to Fort Bragg, North Carolina. Normally on a flight that short the pilot would just fly the radio beam down, but this was an exercise in precision dead reckoning navigation, meaning that the navigator does a lot of figuring, measures wind velocity, keeps an accurate check of airspeed, and without reference to landmarks on the ground, keeps track of his position entirely by computation. It was awfully interesting – I sat down in the navigator's compartment and navigated and never did look at the ground until my watch said we should be over Fort Bragg, and then I looked out the window and there she was. I'd missed it by about a mile – one mile in 200. Then I changed places with the co-pilot and he navigated us back to Langley. Navigation's interesting. On a trip like that the pilot has hardly anything to do; he usually turns on the automatic pilot and then just sits there. And all the co-pilot has to do is watch the gas gauges and switch tanks when one goes dry.

About this army business – I'm a commissioned officer in the Air Corps Reserve, and am on active duty status. I can stay on active duty for a year, and then I can stay longer if I want to or I can go to inactive status – still holding my commission. Or I can take an examination for a commission in the regular army.

I think we've got us a place to stay – four of us are going to rent a house. It's a 6-room house with 3 bedrooms, very nicely furnished and in a nice neighborhood between Hampton and Newport News. The address is 59 Greenbriar St, Hampton, but I don't think we'll be moving for a couple of weeks yet. It has an attic, lined with wallboard, where I'll probably have my transmitter. It's oil heated, with forced circulation, and has a separate automatic oil heater for the water, which keeps water hot all the time. It has an electric refrigerator -- it's got lots of things. A big back yard, too, with a couple of apple trees. It's costing us $75 a month, split four ways. The army, when it doesn't furnish us quarters, adds $40 per month to our pay to rent quarters. . .

That clipping about the airplane crash – more people get killed in just that way than due to any other cause in flying. Inexperienced pilots will try to climb too steeply right after takeoff, let the airplane lose flying speed, and stall and spin into the ground. Experienced pilots will let the airplane roll along the ground until it flies itself off, then climb very slowly, picking up speed at first rather than altitude. That's important because it's speed that keeps you in the air. It's nice to have lots of altitude for cruising and maneuvering, but on takeoff, all you need after getting off the ground is enough to clear the trees and buildings around the field, and you don't even need that until you get to where the trees and buildings are.

8/24/40

Will you send me my high school trig book? It's a dark blue, almost black book, well worn, the lettering on the cover almost gone. The title is "Plane and Spherical Trigonometry" by Bauer and Borrke. I've been reading a little about celestial navigation, and to understand that thoroughly I want to check up on my spherical trig.

If I do get to Los Angeles I'll call up Mrs. Schroeder for you. The trip, incidentally, will take about 15 hours, cruising at about 200 miles per hour, and will use up about 2800 gallons of aviation gas. That's one way, for one B-17 airplane.

8/30/40

There are only four weather conditions that make flying really dangerous; low ceilings (solid overcasts below 1000 feet or so), low visibility near the ground, icing, or extreme turbulence in the air. Most common is low ceiling. Broken clouds near the ground aren't bad; you can go between them. Icing is unusual, and most of the time you can go around it. Besides, deicers, which all big ships carry, are fair protection if you do get into icing conditions. Turbulence is always uncomfortable, but dangerous only when it's severe, as in a thunderstorm.

This afternoon our group had an officers' meeting, at which time it was announced that the group is moving to Puerto Rico on or about October 25. Reserve officers, if we go, will have to agree to stay two years; if we don't want to do that, they won't send us – I think. I'm going to look into the possibilities of transferring to Wright Field, in Ohio, although it's practically dead sure I can't get it. They wouldn't even let us indicate it as a preference when we left Kelly. If I can get Wright, everything will be perfect. If I can't, then I'll either stay at Langley, or resign, or go to Puerto Rico. If I stay at Langley, I'll probably get stuck on some job that will drive me crazy, maybe eventually get transferred someplace worse than Puerto Rico. If I resign -- and they don't like to accept resignations right now – well, I haven't figured out what I might do. Puerto Rico may not be such a bad place in a few months, but for a while I'd be living in temporary barracks. Climate is nice – trade winds keep it fairly constant, and pretty cool. But I just hate to be way out there for two years, even though we would probably fly back to the States every few weeks. I don't know what I'll do, except I'm pretty sure I won't go to Puerto Rico.

9/4/40

Haven't done much flying this week. All but two of our ships are out of commission having work done on them. Yesterday we went up to 20,000 feet for some bombing, but before we got started on the bombing one of the engines started giving us a little trouble so we decided we'd better land and fix it. A part of the carburetor was loose. We could see it in flight -- could see it vibrating, and so we were able to come back before it actually went out of order. We didn't even have to shut down the engine -- came in on all four. Now that I've told you that I suppose you'll be worrying about forced landings all the time. B-17s never, and B-18s hardly ever, have sudden emergency forced landings. On a B-17, three engines have to quit cold before you really have to land and even then if you've got one engine left it's a big help. A B-18 can stay in the air on one engine.

9/10/40 *(following a description of the house he moved to)*

Incidentally, I didn't know it myself until yesterday, but if this house had been here in 1861 (and a couple of others hadn't) we could have looked out our sun parlor windows and watched the Monitor and Merrimac blaze away at each other. That was almost directly offshore from here, out in Hampton Roads. There's a memorial drive along that part of the shore now; it's very pretty. One side of it is residential; on the other side, right along shore, I've driven along at night and seen parked cars so thick you'd think it was a football game. That shows the drive must be pretty, when it's used for a lovers' lane in spite of having houses all along it. . .

You seem surprised at Panama and Hawaii being nice. They are, and the Phillippines are even nicer. Puerto Rico, when it's built up, will be practically a perfect station of its kind; it has all the nice things about the tropics but it doesn't get hot. Temperature is 70-80 degrees all the year round.

9/18/40

Well, I finally got around to visiting Washington. Saturday I took a BT-14 up, came back Sunday afternoon, had a lot of fun, saw Harold. . .

We stopped at the Washington Municipal airport where we watched transports for a while and also saw a passenger sightseeing blimp land and take off. Blimp takeoff is the funniest-looking procedure – the tie-down ropes are loosened, the blimp hangs neutral near the ground, the ground crew all get under the car and gives it a big heave up in the air and then the pilot opens the engines up and takes off. . .

I saw a flying instructor's parody on the 23rd psalm a while ago. It goes:

> My student is a headache I do not want.
> He maketh me to lie down at night very weary.
> He leadeth me beside the high tension wires.
> He anointeth my head with gray hairs; my cup runneth over.
> Yea, though I fly only on clear and unlimited days, I shall fear much evil, for he is with me.

9/25/40

I haven't mentioned Puerto Rico lately because there hasn't been anything to mention. One thing we've found out, though, is that we won't have anything to say about it. They'd originally intended to ask all reserve officers who were going to sign up for two years, but it seems now they'll just send them whether they sign up or not. We do know that not all the officers now in the group will go, but we don't know by name yet anyone who'll be going. Wright Field is still a pretty slim chance, because mostly they don't take reserve officers there. Well, I don't care so much if they do send me to Puerto Rico. At least it will be permanent for two years, which is something I haven't had lately. And we will be making trips back and forth quite often, no doubt. With B-18A's. We're losing our B-17's any day now. They're all going to the West Coast, not only ours but all the others in the country. I don't know why – some people seem to think it's for resale to England.

9/27/40
Dear folks—

I forgot to ask you if you'd send me my army short overcoat and one of my white sweaters. . . From what I've heard, I think I may have use for both of those even if I do go to Puerto Rico.

Last night I was up at 20,000 feet for about 3 hours; first time I'd been up that high at night. Very pretty; I could see Richmond. Temperature outside was 15 degrees F, but inside we had the cabin heater on and were sitting in our shirtsleeves very comfortably.

10/1/40 *(orders to Puerto Rico)*

Well, I guess I'm going to Puerto Rico. Orders are out today. It won't be bad though. They've transferred me from the 10th squadron to group headquarters and made me group communications officer, and that will be interesting… Only I hate to leave my old squadron. Captain Harvey is about the best squadron commander in the army, I think, besides being one of the best pilots. I don't know how much flying I'll get a chance to do now, but it will probably be less than I've done lately. This last month I flew 40 hours, the most I've ever flown in one month before."

10/5/40 *(chain letter agreement about letter explaining marriage)*

That two-year stretch you were asking about means two years in Puerto Rico. Duty outside continental U.S. is always for two years; it costs the government too much to send people back and forth for a few months. As far as the extra and unexpected time is concerned, I think that with the draft going through, reserve officers are going to be pretty much in demand for some time to come. They'll need them to train recruits. Also, in my opinion, as long as there's the slightest chance of our getting involved in a war, it's to everyone's personal advantage to get just as much training and experience as he can. That's life insurance, both for himself in case of war, and to the country as a help in staying out. . .

A couple of days ago I got a wedding announcement from Walter. He married a New Jersey girl last weekend. That was a complete surprise. Some time ago, in the chain letter, some of the fellows were speculating on who in our gang would be the first to get married; I don't remember much about it now, but it was suggested by someone and agreed to, I think, by everyone, that whoever it was would write an essay, not less than 500 words, all about it, the essay to be written on the backs of canned pork-and-beans labels. But that was the biggest surprise I've ever had, I think. As long as I've known him, since our freshman year at U.W., as far as I've known, he's ignored women completely. They just didn't exist. Never even talked about them. Well, I wonder what will be in the next chain letter. . .

Yesterday I was practicing landings in a B-18A for the first time. Normally a co-pilot has nothing to do with the flight controls on landings and takeoffs; he's too busy with other things. But eventually a co-pilot becomes a pilot, and first he's got to learn to take off and land. I'll be eligible for that when I've got about 55 more hours. It's like learning to fly all over again. Flying these big ships is easy, but setting them down – it's lots of fun, really.

Bedtime,

73

Aldro

10/9/40

So, you're going through canning season. Well – don't work too hard. . .

My total flying time now is about 350 hours. And I used to think anyone with 200 hours knew all there was to know.

I don't know when I'll leave for Puerto Rico – either Oct 25 or Nov. 1, I don't know which. My address will be

25th Bombardment Group
Borinquen Field, Puerto Rico

Dupory just showed up with a copy of Anne Lindbergh's "North to the Orient." I've been wanting for years to read that. So now I'll wind this up and read a chapter or two before going to bed.

73 –

Aldro

10/24/40.

I've been running in circles the past week trying to get things done, but I've been enjoying it. I've been calibrating the radio compass on the group commander's airplane for the trip to Puerto Rico, and been trying to persuade the pilot on the ship I'll be on that ours ought to be calibrated too. . . He used to be a communications officer once too, so it's not hard to talk to him about it. I think I'm kind of lucky in that respect – most of the high-ranking officers in the group are old communications men, including the group commander, and it helps make my job easier at times. . .

One afternoon last week Dupory and I went up to Williamsburg to see the sights we'd heard so much about – restored building and museums – pretty interesting. We took pictures of ourselves in the stocks and pillory in the old "public gaol."

10/30/40

So, Elizabeth has a scholarship. I'd like to know more about it; it sounds fine.

I didn't know Miss Green had gotten married. I wonder if she remembers me; I had her for typing, you know. And Chaucer! Haven't they outgrown yet the habit of inflicting him on high school English classes? I guess he's like Virgil; he'll always be there. Only to get Virgil you have to be punch-drunk enough to elect it, but there's no defense against Chaucer. (Why hello, Miss Wilson!). I'll bet Miss Wilson is disappointed in my career; she always wanted me to be a poet or author or some kind of artistic worker. She was disappointed enough when I started to "waste my talents" on engineering, but flying! An aerial truck-driver! I wonder what she thinks about that.

I don't know an awful lot about the swimming situation in Puerto Rico. They have to use nets – shark nets – for ocean swimming, I understand, although the sharks normally stay about 100 yards offshore. They enclose an area with these nets, then set off enough dynamite inside the roped off area to kill anything that's in there, and then open it up for swimming. The barracuda is a smaller fish than the shark, but much faster and more vicious. He has protruding teeth, very sharp, and will grab a chunk out of anything he sees just out of meanness. Besides those fish, there are other things – sting rays, jellyfish, and a few other things – that can give uncomfortable stings but aren't dangerous. They have those around here too, and on the Gulf coast of Texas. I never heard of anything like that in New Jersey, though. I hear they're putting up shark nets now at Borinquen, and I think they'll have an inland pool, too before long.

We take off Friday morning, if the weather's OK, which I think it will be. We'll spend Friday night at Miami, Saturday night at Camaguey, about in the middle of Cuba, and should be at Borinquen Field shortly after noon Sunday.

73

Aldro

CHAPTER 5
Letters from Puerto Rico
November 1940 – July 1943

I used to wonder why Dad was stationed in Puerto Rico during World War II. A summary of his career that he wrote in 1956 at Grandma's request answered my question. "In Puerto Rico I was Communications Officer first of the 25th Bombardment Group, then the 40th Bombardment Group, which had squadrons or parts of squadrons in Jamaica, Camaguey, Cuba, St.Croix, Antigua, Puerto Rico, St. Lucia, Trinidad, Curacao, Aruba, British Guiana, and Dutch Guiana. Our job was anti-submarine patrolling and convoy escort, and my part of that job was to see that we had adequate radio communications."

He frequently visited many of the islands, where he purchased the Haiti tables and other items we grew up with. Among notable descriptions were the delights of using a "water telescope" (i.e. face mask), snow-capped mountains at the equator, driving on the left side of the road, the Dominican dictator whose 13-year-old son was a 4-star general, and the Haitian leader who would demonstrate his soldiers' discipline by marching them over a cliff.

To say that Dad was not enthralled with Puerto Rico would be an understatement. A typical comment: "The P.R. telephone company is predictable only to the extent of being predictably unpredictable."

Commentary on flying included the plane that was too small for him, his first water landing, and the feeling in his first slow roll that "I was flying a different airplane and I didn't know what to do with it. Which, of course, is exactly why pilots are taught acrobatics."

Of note also were two rapid progressions: his promotions from 1st Lieutenant to Lieutenant Colonel in just 2½ years, and his wedding announcement less than 7 months after his first mention of Mom in the 12/15/42 letter.

Underlying it all, of course, was the war, which he noted at least had the virtue of making it easier "to get cooperation among the different agencies of the Army."

11/4/40

We left Langley on schedule at 9 am Friday, stopped overnight at Miami the first night, Camaguey, Cuba the second and got in here at 3:35 EST yesterday (4:35 here). . .

Miami is an awfully nice town – I almost wish I hadn't seen it just before coming here. The weather is good, the town is clean and nice – the whole city has a characteristic smell something like a bakery, but sweeter. After leaving Miami, we followed the overseas highway to Key West, then across to Havana. I was surprised to see that the Gulf Stream shows up so distinctly – almost like a row of breakers, and just as definite. Cuba is awfully pretty from the air – cane fields and palms all over the place . . .

In Camaguey, we stayed at a hotel run by an American – pretty much a gyp joint, but the guy could speak and fully understand English, which made the otherwise exorbitant rates partly justified. I paid $1.50 for a room I could have had for a dollar in the States, and $1.15 for a steak dinner that shouldn't have cost over 60 cents. We were brought in to town from the airport in cars on a road so full of holes we averaged about 5 mph; it wound in and out and especially up and down, through wild coffee patches, past native huts built of fruit crates and thatched with palm fringes and overrun with naked kids. Everyone gaped at us all the time; we walked around the town that night and a regular procession followed us around, and all the people living along the streets hung out the windows and stared at us. . . We walked around the park and every time we stopped for a couple of minutes we'd have to break up a crowd of about a hundred people before we could go on. . .

The classes of people are interesting there. At the bottom are what seem to be the native Cubans – white-skinned, but a very dark white. Then there are the negroes – an awful lot of them seem very intelligent, and quite a few speak good English. And then the whites – Spaniards; they run things. I saw fewer whites who could speak English than negroes who could, though.

11/7/40 – *(dated 10/7/40 – obviously an error as he didn't arrive in Puerto Rico until 11/3)*

Yesterday I hitchhiked a flight in to San Juan to get my car; drove it back in 3 ½ hours for the 90-mile trip. The road is as smooth and wide as most highways in Virginia, but twists around and up and down so much, and is so congested with native pedestrians who don't care if they get killed or not -- lots of them look as if they'd be better off -- that you can't drive very fast. . . The roads down here are filled with commercially operated station wagons called "publicos"; I don't know if they're buses or taxicabs or what, but they're all over and are pretty wild. It's just a case of the man that has the loudest horn wins; some of the cars have half a dozen horns and they're going all the time. . .

This afternoon for the first time I went down to the beach to look around. It's a job to get up and down the bluff (200 feet), but at the foot of it is a very pretty coconut grove going right down to the water. There's no beach, just coral, although I heard there's a beach farther down. Can't go swimming anyway. No sharks, they say; the barracuda chase them all away.

11/17/40

I went in to San Juan yesterday afternoon; we flew in, spent a couple of hours looking around then flew back. We'd brought quite a few officers in with us, but they stayed and will come back today. It's a regular thing around here to send a B-18 or several to San Juan every week-end for anyone who wants to go in. It's called the "Playboy Special."

We all try to pick up the language around here, and can get along after a fashion. The PX sells small Spanish texts, and between them, the natives, and the movies, we pick up a little at a time – very little. .

We've been building a darkroom. The building we're in has an unfinished concrete vault that we can use, and it will work out pretty well. I think we can begin to make pictures tomorrow night.

73

Aldro

12/1/40

Dear folks—

I've been Christmas shopping this week; Wednesday I was at Saint Thomas and yesterday afternoon I went into Mayaguez. Saint Thomas is a pretty interesting place. Most of the people speak English and a large percentage of it is very understandable English. It's very much a tourist town. Cars go on the left side of the streets in Saint Thomas. The place is full of souvenir peddlers and their souvenirs are pretty much all alike. Natives lead burros around the streets, all decorated up with hats with ear-holes, flowers, homemade saddles that are made of bright-colored flowered cloth and look like they belonged in someone's boudoir, for tourists to ride. Women carry bundles, trays of things, on their heads -- I haven't seen that here but they do it over there. The town is full of shops selling foreign made works of art, curios, and things. . .

The next chance I get to spend a night in the States, even if it's Virginia, I'm going to get a good hotel room, soak in a shower whose temperature I can regulate (not one whose temperature depends on how brightly the sun is shining), eat supper in a restaurant where I can order good food as I want it from a waiter who speaks English. Then I'll rent a car whether I've got any place to go or not, and just drive around and enjoy the civilized drivers and good roads, maybe stop somewhere and go swimming where there aren't any sharks. Then I'll go to an air-conditioned theater where the seats allow reasonable legroom and see a good movie, entirely in English, completely without Spanish newsreels and travel talks. Then I'll go to bed, maybe stopping first at a hot dog stand or soda fountain, and sleep until 5:45 the next morning, when I'll wake up, note with satisfaction that no one is going to shoot off a cannon outside my window to blow me out of bed, and go back to sleep and wake up again when I'm ready. Then I'll get up, take another shower, shave in hot water and without having to stand in line, then go and eat breakfast without first going for a 20-minute march with the squadron.

Guess I shouldn't squawk so much. . .

Starting tomorrow, we're going to have Spanish classes twice a week for the next four months for all officers of the group. Some Puerto Rican officers will be instructors.

12/9/40 *(how to eat a coconut)*

Yesterday I packed your Christmas box; there was some extra space so I went down to the beach with one of the other fellows and got some coconuts to fill it up with. I don't know if they're much good or not; one of them is a little green. They may be OK by the time you get them. Milk from green ones is supposed to be best, but for the meat you have to wait till they're ripe. I've never tasted any coconut milk yet that I liked, either green or ripe, but the meat's not bad. You get the milk out by punching one or more of the three eyes around the base; you'll find one of them punches out easily, the other two not so easily. After you've drained out the milk you can open the shell by cracking it with a hammer, or you can do a neater job by tapping around and around the center with the back of a knife until it splits open. Don't try to saw them; you'll get sawdust all over the meat.

Saturday we had a lecture by Gen. Bradley, the commander of the 13th Wing, the next higher unit of which the 25th Group is a part. Pretty interesting talk; he told us all about the development of Puerto

Rico as concerned the Air Corps. The 13th Wing has just been organized since we've been here, and we all met the general after the lecture. He told us, too, that we're to fly every airplane we've got at least 100 hours a month. That's a lot of flying; I guess maybe I ought to get in on some of that. All I need is another 35 hours to be first pilot on a B-18, but at my present rate it will be months before I get it. . .

I think Spanish won't be too hard. Their method is to start out teaching us a vocabulary so we can get the idea what goes on, then iron out the grammar later. For instance, we use the infinitive form of the verb for everything at first.

12/17/40 *(Dad and Mom weren't the only ones who rushed into marriage.)*

The chain letter came the other day – Walter's in Panama now, without his wife. Doesn't know when she'll be able to get there, either. They were married about 6 weeks before he left – they hurried it up for that reason.

12/25/40

Seems to me the last time I wrote you I said something about not getting enough flying. The next day I found myself scheduled to fly two airplanes at the same time, and since then I've been going night and day. Since I last wrote, about a week ago, I've flown about 30 hours, which used to be an average month's work. I don't know when I last had a chance to spend half an hour in my office. It's only temporary, but for a week or so I'll be doing full flying duty in each of two squadrons, and handling communications evenings (unless I fly then) and during extra minutes taken from lunch hours. Twice now I've missed Spanish class to fly.

12/29/40

It doesn't seem at all like Christmas season here. Lately, in the B-18's, I've been listening a lot to the local broadcast stations, using the radio compass on them to develop an instrument flying technique suited to conditions here. They play a lot of Christmas carols all the time, which seems out of place.

Bedtime.

73

Aldro

January 11th 1941

We're all moved into our new quarters now and are making the best of it, I guess. One good thing we have in all our two-by-four closets is a provision for leaving a light bulb burning continuously to keep our clothes from molding. That's quite a problem here; over in the other place all of our clothes got moldy. When these Puerto Rican electricians wired the place, though, they were in too much of a hurry – no, I guess it couldn't have been that, I've never seen one hurry yet. Anyway, if you jarred the walls a little bit all the lights in the building would go out. Today I found why and fixed it. If I'd called the electrician it would never have been fixed.

1/13/1941

This isn't the time for me to be writing letters -- I'm tired and I'm disgusted -- but if I don't write now, I'll miss the airmail. Today we were all told that by noon tomorrow we were to move out of the crowded quarters we've been living in over to another building that will be worse. The building we're to move into was intended for officers' quarters; now we're living in an operations building. But how anyone can build a place like we're moving into and expect people to live in it is more than I can see. Each room is about 8 by 10 feet with two closets. The closets are built so you can't hang anything in them that you'd want to hang; they're not big enough for a coat hanger. Two beds in the room will just about leave room to open the door; There won't be room for any other furniture. Those are our bachelor officers' quarters. According to law an officer is supposed to have either adequate quarters or extra pay in compensation for not having them, and we don't get either. "Adequate quarters" for a bachelor second lieutenant is defined by law as two rooms and a bath. The trouble is that in fine print, someone put in that commanding officers can declare other quarters suitable. And we've just got the kind of a base commander who's been doing and apparently intends to keep on doing everything he can to keep the morale of everyone under him just as low as it can possibly be kept. I don't think there's a more thoroughly hated man in Puerto Rico. Enough griping. Guess I'm just stuck.

1/24/41

As of yesterday, I have 4155 hours flying time, and it's accumulating at the rate of from 30 to 60 hours per month. I have over 200 hours this fiscal year, which has only half gone; most officers average about 250 hours in a fiscal year. But I'm not checked out yet as a B-18 pilot, because too much of my flying has been administrative, which doesn't give me any chance to do the things you have to do to get checked off. I'm working on it, though.

1/27/41

Just finished a late supper after landing at 6:45 from a six -hour flight. We flew during sunset; sunsets down here are always pretty, but from the air they are more so and this one was still more so. I had my camera along, loaded with color film, and took several pictures that should be good. We were flying along at just about the cloud level, in and out among big cumulus clouds. The sunset itself had the usual reddish colors, and in the east the sky was purple, blue, and kind of pink, with clouds of an unusual sparkling gray standing out against it.

I spent the weekend in San Juan; went in Saturday noon, shopped some, and in the evening I saw "The Great Dictator" --

Charlie Chaplin's show. I was surprised to find it in town; hadn't expected to see it here for a long time.

2/3/41 *(the value of learning to type – I couldn't agree more)*

I don't know how much typing Elizabeth has had – you mentioned her not wanting to go on with it. I think, though, that if she can type well enough then she can use her time to better advantage somewhere else; but if she's not very good yet, then she ought to keep on until she is. I think typing has been as valuable to me – the one semester I had – as any other subject I took in high school except mathematics and physics.

2/7/41

Dear folks—

Today for the first time this week I'm home in time to wash up before supper. It's quite a luxury. A representative day for me is: up at 6:00 in the morning, dress, rush over to the squadron for a 20 minute walk with the men; then over to breakfast and back to quarters to shave and things. Then, about 7:30 or 7:45, over to 10th squadron operations to see if I'm flying in the morning and sit in on any conferences they may have pertaining to flight. Then, if I don't fly, I walk over to my office to see how things are going, take care of correspondence and other paperwork that's always coming up, then spend the rest of the morning working on whatever else I have time for. Dinner at 12:00, then down to the line to take off in Number 12 to fly for six hours with major Hillery. During that time, I fly the airplane most of the time, but don't do much in particular, except maybe get in a little time under the hood or something. We land about dark, at 6:30 or 7:45, and I rush up to the mess to see what's left from supper. About the only way to relax after supper is to go to a show at the post theater, so quite often I do that; the show is over about 8 or 9, and then I go back to my quarters -- 12 to 14 hours after I left. Number 12 is out of commission now, though, getting its brakes replaced, so I'll have some peace for a while."

2/14/41
Dear folks—

This afternoon I've been flying around at 15,000 feet. It's the first time I've been that high around here in the daytime; the most noticeable thing was that reflections in the ocean of clouds near the horizon were nearly perfect, in spite of the rough water. Major Hillary had his dog along; she didn't like the air at that altitude (we were using oxygen, of course) so we dropped down to 10,000. Night before last we went up to 18,000; on our way up at about 11,000 feet we saw the moon come up on the ocean. Prettiest moon-rise I've ever seen.

We listen to the radio a lot when we're flying. Several of the stations have "American Hours," when they play American records. .

I have my big easy chair in my room now. I'm always stumbling over it, but it's worth it having something besides a bed to read in.

Last Sunday a bunch of us went over to Saint Thomas and went swimming and had a lot of fun. . . I don't think I'll go over this weekend, though. I think I'll wait until the sunburn I've got now finishes peeling. You should see it. I was only out a few hours, but the sun down here is a lot more vicious than at home. I had it all over -- back, front, face (you should see the skin hanging from my nose and forehead!) and on my legs too. My legs are still a little sore.

2/18/41

El Morro, in San Juan, is an old Spanish fort used now for barracks, recreation facilities, and a few shops. . .

I've taken a lot of color pictures since I've been here, and I'll send them home for you to see soon; I ordered a projector from the PX some time ago and I want to wait until it gets here so I can see them myself first and include intelligent explanations. I'll want them back, though. I want to see if I can't somehow make black and white prints from them.

73

Aldro

2/23/41 *(promotion to 1st Lieutenant)*

A couple of days ago I was promoted to first lieutenant, with several other reserve officers. . . I'm wearing nice new shiny silver bars instead of the old brass ones, and my clerk gets a kick out of putting "1st Lieut" at the end of my official correspondence. . .

I guess my sunburn must have sounded pretty funny to you with your 10 and 20 degrees below weather. I was just thinking – it must be three years since I've seen the thermometer below zero. In New Jersey it got cold and snowed a lot and froze sometimes, but I don't think it went below zero, and in Texas, of course – well, when it froze and snowed there it crowded the war news out of the headlines, and when it got as low as 20 degrees (there's only one side of zero in Texas) a lot of people didn't see how anyone could survive. And I was one of them.

2/28/41 *(Trinidad)*

Dear folks—

Been feeling kind of worn out yesterday and today. I got back yesterday from Trinidad -- we'd gone down Tuesday, spent Wednesday looking around, and came back Thursday. Very nice trip, and we had a lot of fun, except for getting involved in a party that lasted until 2:00 AM one night. Trinidad's a nice place at this time of year; cleaner than Puerto Rico, and towns are much more intelligently laid out. In Port of Spain, the main town on the island, sidewalks are wide enough for three people to walk abreast and two cars can pass on the street with cars parked on one side. In Puerto Rico such wide open spaces are unheard of. Everyone down there speaks English, but accents are pretty varied, ranging from the genuine British to an accent sounding a lot like Spanish. But anyway it's all English.

3/8/41 (*"water telescopes"*)

Dear folks—

Hasn't much been happening this week; nothing to write about. I think the rainy season is starting -- we've had one or two showers a day the last few days, and the island is more generally covered with clouds than it has been. Major Hillery and I went up to 20,000 feet the other day to look the situation over -- you can really see just what's happening if you can get high enough, and 20,000 was high enough that day. Apparently a south wind, more moist than usual, was blowing and as it was lifted over the mountain range running the length of the island, the moisture would condense into big, turbulent cumulus clouds. It looked like surf breaking over a reef. The clouds themselves had just that appearance.

It's quite a job getting a B-18 to go up that high. B-17's will do it easily and a lot higher, but the B-18 is more heavily loaded for the power it has, and isn't as well supercharged either. By the time we hit 20,000, the airplane was just staggering along, and if I let go of the wheel at all she'd start down right away. I felt like I was lifting myself by my bootstraps.

I had to laugh at myself -- I was trying to whistle and couldn't. Not enough air. It wasn't that I couldn't blow hard enough -- I was using oxygen, of course, -- but what I blew wouldn't whistle. I don't know exactly why, although it undoubtedly is connected with the rarified air. . .

I was over to Major Hillery's house the other night for supper – had fried chicken and corn on the cob. Mrs. Hillery is from Virginia and prides herself on her fried chicken, and it is good, too; but the corn on the cob! They don't raise any vegetables on the island, you know – nothing but sugar cane. All our lettuce, carrots, cucumbers, and everything is all imported. But it seems a captain here, a Puerto Rican, has a farm near San Juan where he raises some things and he'd given them the corn, and they asked me over and it was really good.

Chain letter's here. Walter's wife is with him now, in Panama. Wayne and Alan and Ketchum write about skiing and skating – makes me homesick, but I can retaliate to some extent by writing about swimming. It's really nice over at St. Thomas. A lot of the fellows have water telescopes – a kind of water-tight transparent mask you can wear which makes it much easier to see under water. You can swim around under the surface looking at rocks, and reefs, and holes, and see all kinds of fish and things, brilliantly colored. I spent about half an hour one day trying to catch a little bright blue-and-gold fish. He'd swim into a hole and I'd corner him – just a small hole, big enough for my hand – and I'd reach in after him and he'd slip by me every time. He was a very sportsmanlike fish, though. Every time I had to come up for air he'd stand still, so I could find him again when I came back down. Finally three of us started trying for him and then we lost him.

I forgot to tell you about the Chinese restaurant we ate in one night in Trinidad. I've eaten in Chinese places lots of times . . . but at this place I had my first experience with chopsticks. They gave us regular silverware with our food, but the two English officers we were with asked for chopsticks so I did too, with the idea of trying them out. But when they brought the chopsticks, when I wasn't looking, the waiter took away my silverware so chopsticks were all I had and I had to get along with them. Spaghetti's not hard to eat with them, or chow mein, or chop suey, but then they brought out some stuff that had to be cut, and that wasn't so easy. But it was interesting.

The other day I went through one of Pan-American's Clippers in San Juan. Four 1500-horse engines, room for 60 passengers or less, depending on gas load, regular luxury Pullman bunks – quite a boat. And the navigator's room – it's bigger than my bedroom. Some stuff.

73

Aldro

3/19/41

Last week our group got the new equipment the radio station's been waiting so long for – a 400-watt transmitter mounted in a specially-built truck with its own power supply, and a receiving station with remote control equipment for the transmitter all installed in a trailer. It's a beautiful piece of equipment and it's why I haven't written. Just about every minute when I could be on the ground I've been working with that new equipment. I'm awfully mad, though, about what's going to happen on April 1. Our group splits, forming two groups, the 25th and the 40th. The 25th will keep the new radio and I go with the 40th. I waited six months for that equipment and now three weeks after I get it they transfer me to another group. I'll still be group communications, for the 40th, but we won't have a thing -- not even any airplanes -- for no telling how long.

3/22/41

It's rumored that the 40th Group will get B-26's, a twin-engine ship that's supposed to be faster than any pursuit ships now flying in

Europe. That's news-talk and to be taken with several pounds of salt, but from all I can gather, they really are a good airplane. . .

They checked me out as first pilot on a B-18 some time ago, but I hardly ever fly except as Major Hillery's co-pilot. This group headquarters job is a pain in the neck in some ways, but I still wouldn't trade jobs with anyone in the group.

By the end of this month I'll have 600 hours' flying, and in two months I'll be eligible to check out on a B-17, if we had any B-17's and if I was going to be in the group that's going to get them if they send any down here.

3/25/41

Dear Margaret –

Biggest news around here is that General Daley, the commanding general of the Puerto Rican Department, leaves tomorrow – permanently. Amid much rejoicing. He's the man who to date has required all military personnel to take a 20-minute walk with the troops every morning at 6:25, parades twice a week, curfew at midnight every night, regardless, -- and the one thing I don't know if I've mentioned – all officers carry swagger sticks – canes. You should see me and my cane. It's a very light piece of bamboo, just long enough to reach the ground, thick enough to hang together – it complies with the law, and that's about all.

4/4/41 *(baby alligator)*

Do you want me to send you a baby alligator the next time I go to Miami? I would have sent you one this time, but I wanted to ask you first. All I'd have to do would be to go into one of the stores there and give them your address and they'd take care of everything, and they guarantee that the alligator will get there in good shape.

4/11/41

It's too bad you don't want an alligator, because I feel like sending one to someone, and I don't particularly want one myself. They're not big enough to be dangerous, you know. They're pets.

4/20/41

Why do aviation movies have to be so full of bunk? . . . I think in every aviation picture I've ever seen, there's been at least one scene where, for some reason, a pilot loses control of his ship, usually by losing consciousness for lack of oxygen or something, and the ship immediately goes into a vertical power dive, straight for the ground at 500 MPH or so, and finally the pilot becomes conscious and by a superhuman effort pulls out of the dive just in time to miss the treetops by inches, maybe picking up a few leaves on his undercarriage. Or else he crashes. Anyway, it's the silliest thing imaginable – the idea that an airplane goes into a dive as soon as you relax on the controls, and the idea that, once in a dive, she'll stay there and build up all that speed. As far as I know, every airplane that's been built in the last ten years and I don't know how long before has had controllable trim tabs . . . so that the pilot, no matter how the ship is loaded, can adjust it so she'll fly straight and level hands off. I'd certainly get awfully tired if I had to hold an airplane from going into a dive for six hours or so at a time. And as for staying in a dive – every airplane tends to nose up as it gains speed, and I doubt very much if I have enough strength in my body to hold a B-18 in any such dive as the movies show ships doing by themselves.

4/25/41

Myers and I both went in to San Juan this afternoon, but instead of taking a B-18, we took the two A-17's and flew formation most of the way. It was fun. I haven't done much formation in single engine ships in the last year. . .

Yesterday we flew a review for the new base commander – General Bradley. It was the best review and the most snappily executed I've seen.

5/8/41 *(about a trip to Jamaica)*

Guess I'm kind of delinquent in my correspondence. Sunday night I'd just finished unpacking my bag after getting back from Miami when someone came along and wanted me to go to Jamaica in the morning, so I had to pack up again.

Captain Cecil, the weather officer here, was the pilot. It made the flight a lot more interesting, having him there to explain the weather as we came to it -- all pilots, of course, are supposed to know something about weather, but -- well, there's a saying as follows: -- "Pilots are people who know less and less about more and more, until finally they know nothing about everything. Weather men know more and more about less and less until they know all about nothing. Radiomen start out knowing all about everything, but due to association with pilots and weather men, finally know nothing about anything."

There are an awful lot of things pilots are supposed to know a lot about -- weather, engines, radio navigation, aerodynamics, -- all kinds of things. I don't know nearly as much about weather as I'd like to. I saw one thing I'd never seen before --cloud formations flowing down valleys over in Haiti.

I should know the way to Jamaica pretty well now. I went in that general direction Friday to Havana --it was out of my way, but I wanted to avoid some weather that was reported from farther north. Jamaica is nearly due west of here, about 800 miles. The route goes along the south shore of the island of Santa Domingo; mountains over there go up to 10,000 feet in some places, and it's very pretty, especially where they drop right down to the water. Can't take good pictures of it though.

The hotel in Kingston is awfully nice. The rate for rooms includes meals and what meals! I'm sending you a menu separately. Monday afternoon we went swimming in the hotel's outdoor pool and were served tea on the pool terrace! While we were having our tea an old Englishman in trunks, jacket, slippers and monocle walked out to the pool, took off his jacket and slippers, walked up to the edge of the pool, checked the fit of his monocle and dove in, monocle and all. He swam around that way for about 15 minutes. Amazin'! . . . I ate mangos in Kingston for the first time. Papaya, which they call paupaw, I'd had before, but had it again. It's good. So are mangoes. Bananas they serve in lots of ways. I've eaten them fried in San Juan (fried plantain), and in Kingston we had some baked plantain. The plantain is a big overgrown banana, woody in texture, but very good when it's cooked. Besides ordinary bananas, they have some small ones here to about 1/3 to 1/2 normal size, thin-skinned, very sweet. They're called niño, which is the Spanish word for little boy.

Saturday morning I have to give a lecture to the navigation students on radio direction finding, and this weekend I'm going into San Juan. Krisberg and I are going up to El Yunque, the highest mountain on the island. It's on the map I sent you -- the Caribbean National Forest, on the east end of the island. The top of the mountain is nearly always covered with clouds, and the jungle is real tropical jungle. From the air you can see giant ferns, thick underbrush, and they say there are seashore crabs in there and oysters growing on tree-trunks. I think it's about time I was seeing some of Puerto Rico on the ground.

73

Aldro

5/16/41

Sunday, I went into San Juan and Plummer (civilian engineer) and Krisberg (Signal Corps officer) and I went mountain climbing. We drove as far as we could and hiked the rest of the way along a well-cut trail through the jungle to the top of Mt. Britton, one of the four highest peaks on the island. It's a pretty trail; the jungle has all kinds of things growing in it - giant ferns, mahogany trees, and is full of what seemed to be wild orchids. It rains a lot up there and the mountain streams and waterfalls are very pretty. . .

Back at Langley I used to do well to fly 30 hours a month. The month of May is now half gone and already I have 50 hours.

5/24/41

The other afternoon I took an emergency hospital case into San Juan -- appendicitis. The last time I was on one of those trips, they sent a sedan out to get the man instead of an ambulance. And this time they didn't send anything at all. I was certainly burned up. I don't think the man was very bad, though – we'd gone through some pretty rough air on the way in, and had to circle the field for a while waiting for a heavy rain shower to get out of our way so we could land, and he stood it all OK. Better than I would have. He was on a stretcher in the back of the ship, and I never can ride back there without getting sick.

5/30/41 *(very unflattering comment about Puerto Rico)*

I can't get leave. They're giving only emergency leaves, or leaves to get married. A lot of the fellows are doing that now. I think it's pretty foolish to bring a wife down here, but I guess it would be just as bad to wait. This is no place for white people. It was certainly a misfortune for humanity when this particular mountaintop stuck it's head up from under the ocean.

6/17/41

I'm teaching school. I have six hours to tell the new co-pilots all about Air Corps communication. It's not easy. . .

I saw "I Wanted Wings" in Miami. You'll have to see it. It's the story of four flying cadets starting out together and what happens to them as cadets and officers. It's pretty realistic, but too highly romanticized in some ways. One detail amused me a lot: a B-17B has brakes with a very characteristic squeak. The movie reproduced the squeak perfectly, but on a YB-17 instead of the airplane that actually has it. You have to look close, though, to know the difference between the two airplanes from the outside. The flying in the picture is somewhat spectacularized, but not too inaccurate.

6/22/41 *(about flying acrobatics)*

I'll bet you enjoyed "I Wanted Wings". . .

I was just thinking about "I Wanted Wings" – the scene where the cadet falls out of the airplane – and some of the things it reminded me of. My first slow roll was about the most puzzling experience I had – my feet fell up off the rudder pedals, things in the cockpit flew up and dangled in front of my face – worst of all, once I got over on my back it seemed like I was flying a different airplane and I didn't know what to do with it. Which, of course, is exactly why pilots are taught acrobatics.

A classmate of mine fell out of his airplane at Randolph – Charlie Martin. He got in a spin and couldn't get out, so he opened his canopy (flying solo), unfastened his belt, then tried once more, before bailing out, to break the spin and succeeded – and got pitched out of the cockpit in doing it. He landed OK with his parachute, and the amazing thing was that his airplane landed OK too – all by itself, without a pilot, in some farmer's pasture. The airplane was damaged less than the pilot -- he got a shaking up and a lifetime of kidding. . .

In airplanes with retractable landing gear there's an arrangement so that if you cut the throttles back with your wheels up a loud horn blows in the cockpit. That's to keep you from forgetting to put your wheels down to land. Yesterday coming back from St. Thomas, Captain Decker, of the Coast Artillery, was sound asleep in the navigator's seat right behind the pilot. We decided to have some fun. Myers pulled the throttles back, the engine slowed up, and that horn cut loose with a long, loud, blast. Decker's eyes popped open like headlights being switched on – he started clawing frantically for his parachute – (foolish thing to do – we were over water and a parachute would have drowned him; better to stay with the ship in case of a forced landing) – then he saw the rest of us laughing and went back to sleep.

Those stories go on without end.

6/30/41 *(Dad has described Puerto Rico weather as less hot than Miami and even pleasant. I remember Puerto Rico as being very hot, even in December, when we always visited.)*

I've mentioned the weather we have here. The reason it doesn't get hot here is because the wind keeps it cool – we always have from a 10 to 25-mile-an-hour wind here.

7/7/41

I had a few bad moments yesterday at Saint Thomas. I was coming in for a landing in a pretty strong wind. Over there, the mountains make the wind do funny things; in this case, I suddenly found myself, while I was still about 100 feet off the ground, flying into 20 mph less wind than I'd just had, which meant I lost 20 miles an hour airspeed suddenly and the airplane naturally dropped. I was able to pick it up just in time with the throttles -- I had visions, though, of the landing Wilder made at San Juan a month ago. He landed too soon, hooked his tail wheel over a log lying across the end of the runway, and ripped the tail of the airplane up pretty much. The area where I'd have landed if I hadn't been able to gain speed in time was under construction, and I'd probably have broken my landing gear too. We had something like that happen right here at Borinquen a while ago -- one of the boys in the 27th was trying to see how close to the end of the runway he could land and got too close and pulled his tail wheel off on a bank of dirt. No one ever gets hurt when things like that happen, but it takes time to repair the airplane, costs the government money, and is very, very tough on a pilot's ego. But the same pilot never does anything like that twice. I'll never go into Saint Thomas at less than 100 miles an hour again.

73

Aldro

7/26/41 (*Only mention of bombing so far. Flying usually seems to be to get to a place for work or just because someone wants to go there.)*

Monday night in Miami we went to an indoor ice-skating rink I never knew about before and had a lot of fun. I didn't like the skates they had, though; will you send me mine? I want to go there again the next time I'm in Miami.

Today I got orders transferring me to the Wing and making me Wing communications officer. I guess maybe Air Corps organization is kind of confusing to you – the lowest tactical unit is the squadron. Several squadrons form a group, and several groups form a wing. In Puerto Rico we have the 13th Composite Wing; in command of it is General Bradley. He's leaving in a couple of weeks, though, and we don't like to see him go. . . Several weeks ago General Bradley called me up to see him and we talked about the possibility of giving me that job (so far no one's had it), and he asked me whom other than myself I thought could handle it and I suggested Ambrose, who has my old job as 25th Group Communications. So, the General interviewed Ambrose too, and then asked him who besides himself could handle the job and Ambrose said me. So, there we were; the only two people under consideration, each of us had recommended the other without the other's knowing about it. We had a good laugh over it later when we found out what had happened.

I'll like the job I know. . .

I have a room to myself now, without a room-mate. . .

This morning I flew a bombing mission at 18,000 feet. Bombing is about the most exacting kind of flying there is, but it's fun for just that reason. I won't be doing much of that now, I guess.

8/1/41

Dear folks—

Today, one week after the date of the order putting me on the wing staff, I lost my Wing Commander. . . Everyone's really sorry to see him go. We've been particularly unfortunate here in the commanding officers we've had, and General Bradley has been the one exception.

I was over to St. Croix yesterday and went into town for the first time. There are two towns over there, Christiansted, where I went, and Fredericksted. The island is pretty much like St. Thomas; population mostly negroes, with the whites predominantly Danish. (Over at St. Thomas today I noticed a sign on the runway of the field saying "KEEP OFF THE RUNVAY"). The British influence is less noticeable at St. Croix than at St. Thomas.

Every time I go to the Virgin Islands, I'm fascinated by the hats the people wear, especially the women. The dresses, for one thing, are pretty bright-colored, and they all wear straw hats with bright ribbons for hatbands. It's the same type hat as a man's felt hat, except it's straw, a closely-woven, finely braided straw. . . In Puerto Rico it's the men's hats that are funny. No Sunday suit -- and I do mean <u>no</u> Sunday suit -- is complete without a flat brim, flat crown straw hat. It's a type commonly seen in the States, but the brim is a little narrower and they're worn higher on the head. Also, they're worn with clothes of a type we call winter suits, which makes them more noticeable. The other type hat that's characteristic is what the "jibaro" wears, the plain ordinary goof. They all wear straw hats too; the hat is woven in much the same way as a basket, except that it has a brim, a brim wide enough to be useful. And apparently in making those hats they just weave until the brim is wide enough and then stop, leaving unfinished ends sticking out about two to four inches like spokes all around the brim.

8/3/41

My new job isn't new in that I have other communications officers under me – I had that before. One new thing, though, is that I have to fly around the islands to the different fields to check on how things are going. . .

Yesterday they installed a couple of electric water heaters in our building. They haven't got the power connected to them yet, but that won't take long and by tomorrow we ought to have hot water.

73

Aldro

8/15/41

Dear Palefaces and Elizabeth –

I don't remember when I wrote last; I've been in a perpetual storm for the past few days, with all the stuff that goes on in the office and flying besides. Incidentally, you can tell Aunt Johanna something that will comfort her; any flying I do now is purely coincidental. In addition to the job I have now (which is supposed to be handled by a major with two captains as assistants) there are attempts afoot to talk the powers that be into shoving off a couple of other high-powered jobs on me that I definitely don't want, although they do touch remotely on communications.

At Monmouth I used to gripe about sitting around doing nothing all day. Remember?. . .

$700 for uniforms for a first lieutenant includes a lot of formal dress uniforms I wouldn't be seen in, like the ones I stayed home from Peru rather than wear.

My ice skates came today; thanks for sending them.

We have running hot water here now; it's certainly nice to shave every day in hot water instead of only when we go to Miami, and to have hot, warm, or cold showers by choice.

9/22/41 *(Haiti, and extreme inhumanity)*

Last week-end I spent in Port-au-Prince, Haiti. It's the capital of Haiti, and the only town in the country of any size. It's a 2:20-hour flight from here direct, over some very pretty country. Pretty to fly over, that is, but some of it would be pretty bad for a forced landing. Some of the roughest mountains between here and there that I've seen anywhere.

The language in Haiti is French, but almost everyone can speak some English – it's better in that respect than P.R. Port au Prince is a clean town, and not at all crowded like most towns in the West Indies. Except for the difference in architecture, the streets are very much like in the States and in lots of places you can make U turns in the middle of the block. The streets are very clean, in spite of the large number of little donkeys all over the place carrying loads twice their own weight.

I've never seen people carry things on their heads as universally as there. Here it's very rare; in the Virgin Islands, Jamaica, Antigua and Trinidad it's done, but nowhere like in Haiti. Bundles of laundry, baskets or dishpans full of groceries, bundles of long reeds, furniture – everything imaginable. . .

Saturday night we visited a couple of voodoo dances. They have them right in the city. . . They have a few candles lighting the dance floor, and usually for an orchestra a couple of tom- toms and a man making some kind of noise through a bamboo tube several feet long. Everyone stamps the floor and sways and chants to the beat of the tom-toms. One little kid about 10 years old came over to where we were standing and put on a performance of his own -- I've never, anywhere, seen such energetically wiggled hips. . .

One of the US's newest battleships was anchored in the harbor so we went out and looked her over. First time I'd been on board one. Sunday when we took off we buzzed her, then headed off into the hills to look for an old Haitian fortress called "The Citadel." It's built on top of a steep 3000-foot mountain. It belonged to a man who had set himself up as a local emperor or something, a long time ago. He had an army whose discipline was his pride and joy; he used to amuse himself and show off to visiting generals by marching them over the cliff -- a 3000-foot drop. He'd line them up facing the cliff, command forward march and let them go over the side; then he'd command halt and whoever was left would halt. The place is in ruins now. We flew close to it and took a lot of pictures.

Sunday

This morning Collins and I were down on the beach and had some target practice. I'd borrowed a .22 pistol and he had a .22 rifle. We'd throw bottles out in the surf and shoot at them (no one swims there). I got a kick out of the fact that, at 50 to 75 feet, I could usually break the bottles with the pistol before he could with his rifle.

One of my jobs now includes commanding a detachment of men, with all the attendant personnel paperwork, concern over quarters, mess, and personal life of the men. That's a job I never thought I'd have. Usually in the Air Corps a communications man can depend on having nothing but staff jobs, where he makes plans and coordinates communications activities and leaves command functions up to someone else. Yesterday was payday; as commanding officer, paying the detachment was my job. Being a small detachment, the payroll was only a little over $900. I get a kick out of some parts of that job, but I'm afraid there'll be a lot of headaches too.

73

Aldro

10/13/41

Dear folks—

I've been having quite an eventful time lately. Let's see – last Monday – a week ago tonight – we had a blackout of the whole island. Cities and towns have been having individual blackouts as long as I've been here -- I think I told you about a couple of them that I flew over. Anyway, this time it was the whole island. We had every airplane in the air over different parts of the island, and most of the ships had newspapermen and radio announcers aboard. I'd had the job of arranging the radio end. Every broadcast station on the island had an announcer in one of our airplanes describing the blackout and being rebroadcast. One of the broadcasts was in English and it was really good.

Thursday the Wing Headquarters Squadron had Organization Day. That's a squadron's birthday party. All the officers had dinner at the squadron mess, and in the afternoon a softball game (ouch!) . . . The General was there – he didn't play – but old Col. Ignico did – he's a big fat man, second-highest ranking officer on the field, a grouch most of the time, but on Organization Day he gets drafted to play ball and he plays.

The worst of it for me was that I was to make a night flight to Miami that night and ought to have been in my quarters getting some sleep.

Takeoff was scheduled at midnight, but the airplane developed some last-minute trouble and we didn't get off until 1:00 AM. One landing light wouldn't work, and I wouldn't take off on a long night trip without both of them working. . .The trip itself was fine. Moonlight all the way in good weather. At 9000 feet, I was above all the clouds all the way. The last sunrise I'd seen from the air was nearly a year ago, but no sunrise was ever as welcome as this one after a night of fighting to keep awake. . . Once the sun came up I wasn't sleepy any more. We got to Miami about 6:30 AM, and when we got there they wouldn't let us land. Ground fog. . .The fog layer was only about 100 feet high, and although horizontal visibility on the ground was reported as only an eighth of a mile, from above it looked almost clear. So I got permission from the tower to buzz the field to see how bad it really was, and when I did I found my prop wash blew the fog away so I went around again and landed. Funniest thing I ever saw. A thicker layer of fog wouldn't have done that, of course, but in this case all the air needed was a little stirring up and the fog dissipated. Those things are pretty interesting. . .

Did I tell you we got a P-36 (pursuit airplane) assigned to Wing HQ? I checked off on it last week. . . I certainly enjoyed it. It's a single-seat, low-wing monoplane, closed cockpit, 1000 HP engine, cruises about 200 MPH but is good for 300 if you push her. She's got more pep than any ship I've ever flown. I didn't do any acrobatics my first time up – hardly enough time. Bomber pilots need to fly an airplane like that once in a while, just to blow off steam, although for a permanent assignment I much prefer a bigger airplane that I can go places in, and fly through bad weather if I have to, and land in a cross-wind without ground-looping.

Braddock and one of the other fellows each bought a new Philco combination radio-phonograph with automatic record changer in Miami this week-end, and they keep needing advice about breaking them in. So I'm busy.

73

Aldro

10/28 *(Dad took us to see the "Ice Capades" when we lived in Germany)*

Dear folks—

I got back from the states Sunday after being away 12 days. The purpose of the trip was to inspect some of the communication facilities along the East Coast. Spent one night in Tampa, two in Washington, 5 in New York, one at Langley and two in Miami.

In New York. . . I saw the "Ice Capades" at the Center Theater, where the stage is a skating rink. . .

I spent a couple of afternoons in Central Park taking pictures. It's an area of about 5 square miles, right in the center of Manhattan, surrounded on all sides by skyscrapers. There's a lake, with boats you can rent, bicycle-, bridle-, and footpaths, woods, streams, stables where you can get horses – it's everything you'd least expect to find in the center of Manhattan. . .

Walter is going to St. Lucia, an island about 400 miles southeast of here. A British possession. I'll probably see him now. He has to send his wife back to the States.

73

Aldro

11/5/41

My new address is – HQ 6th Interceptor Command, Borinquen etc. . . The 6th Interceptor Command was formed, took over Wing HQ as its own HQ. The headquarters here is in command of all pursuit aviation in the Caribbean, from Panama all the way around, so in a way it's a promotion for all of us. But I'm disappointed. I've spent a year and a half in bombardment and now I'm in an Interceptor Command. All I know about pursuit is what I learned the day I flew a P-36 for half an hour. And in a staff job like this it won't make any difference whether I'm a pursuit pilot or a bombardment pilot; the only airplanes I'll get to fly of either type will be the obsolete ones – I'll never see the inside of a B-25 or P-39, the new ships we're supposed to get.

Our table of organization doesn't even have a communications officer. I'm listed as an assistant operations officer -- we have scads of those. There will be communications work to do eventually, but right now I'd certainly be glad to be back in either the 40th or the 25th Bomb Group. And stay there. I've had enough of higher headquarters.

Did I tell you I got a new enlarger? I bought it through the Post Exchange and it came last Saturday. I made some pictures the other night -- I'll send you some.

11/13/41

Dear folks—

Your letter about Truax came the other day, and today the one about Frances Baldwin. Too bad. I didn't know Truax's family, except his wife.

A pursuit airplane is pretty hard to fly in bad weather on instruments. . . Pursuit ships are usually less stable than heavier ships and it takes more concentration to fly them; things happen faster, and in very rough weather things happen faster than the instruments can follow, which is pretty bad. Besides all that, a pursuit pilot has to do everything himself, -- no co-pilot or navigator to keep track of position, keep an eye on engine instruments, change gas tanks when necessary, and tune the radio as needed. For those reasons the Army normally doesn't allow pursuit ships to fly if there's a possibility of hitting instrument weather.

12/5/41 *(Includes an attempt to write in Spanish)*

Wednesday, I took a trip to Ciudad Trujillo, in the Dominican Republic. . . I saw Christopher Columbus' tomb in an old church on the main plaza; the church was built about 1514, I think, and is supposed to be one of the oldest in the Western Hemisphere. . .

Margarita – es usted segura usted desira "cuantos" para Natividad? Yo no comprende porque cualquien desira "cuantos" para Natividad. O intende usted "panuelos"? Es mas probablemente. (Perdone la gramatica – no hace el diccionario conmigo hoy).

Trujillo City is laid out like any Spanish town, but is much cleaner. It was wiped out by a hurricane and had to be almost entirely rebuilt in 1930; therefore it hasn't had time to get as ramshackle as San Juan. The people make an effort to keep it nice too; the streets are clean (unheard of in Puerto Rico) and buildings neatly painted. . .

I see my being in the Interceptor Command has you worried. Well – I haven't flown a pea-shooter yet, except a half hour in our P-36 before Major Alan ground-looped it. I will get a little time in them, but most of my flying will still be in B-18A's, the old standby. I'm a staff officer, you see, and flying is incidental, and since we have a B-18A, that's what I fly. I'm not a tactical pursuit pilot, and no one will try to make me one. If I get away from this staff job to go back to flying, it will be as a bombardment pilot, not pursuit.

12/11/41 *(first mention of war)*

I suppose you're frantic with worry now that war has been declared, and I doubt if anything I told you would make you worry any less. Actually, I don't expect any trouble to develop here, although if it does we won't be caught unprepared as they were in Hawaii.

12/21/41

Tonight, I'm in San Juan. I hate this town. In a year I've voluntarily spent one night here. I got word at 4 PM this afternoon to come in, prepared to stay several days, so I'm here. . .

I've been trying to transfer to a bomb outfit, but I'm having trouble getting out of this headquarters. Incidentally, I was over at Ponce the other day and checked off in one of their new pursuit ships. It's one of the easiest ships I've ever seen to fly, land, take off, or handle on the ground. . . Only trouble is the airplane's too small for me. I had a hard time getting in the cockpit in the first place, and once in I had to sit all hunched over in order to see. I asked the fellow who checked me off if he didn't have one a couple of sizes larger, but he didn't.

12/26/1941 *(return address: Army Detachment, HQ 10th Naval District, San Juan, P. R.)*

I'd like to say thanks for the Christmas presents but I don't know yet what you sent me. It's still out at Borinquen. I haven't been able to get out there since I came in here all of a sudden last Saturday. . .

Apparently we'll be here for some time; my address is above. All I have with me is the bundle of clean laundry I picked up and took along last Saturday, and that's not all clean any more. I'll probably have to move out of my quarters at Borinquen and into quarters here, which will be no sacrifice (you should see the quarters here!), but there's no telling when I'll be able to get out there to do it. My car is still out there too. Everything but a few spare shirts.

12/28/41

This war has done one good thing. It's made it easier to get cooperation among the different agencies of the Army, and the Navy too. For instance, I can't imagine the Navy letting an Army officer into its BOQ for anything but a visit, before the war. Before the war I used to beg the Signal officer at Borinquen for bits of equipment I needed, fuss with requisitions, tables of allowances, and usually not get anywhere. Now, if I want a $10,000 radio station, I can pick up my phone, call the Department signal officer himself, tell him, "Colonel, I need a radio station; it will cost $10,000," and he'll say "Just a minute" and I'll hear him say to someone in his office "Radio Washington to send complete equipment for this kind of a radio station, right away." And that's that. Such a system is sure to lead to unnecessary expenditures for lots of things, of course -- a careful investigation beforehand would show better and cheaper ways of getting the desired result -- but in war the situation changes and any action right away is likely to be better than a carefully considered action six months from now.

1/9/42

Your description of the snow and the cold spell make me homesick. I don't know why the uncomfortable things should have that effect so much more noticeably than anything else; certainly I can remember pleasanter things than having to walk two miles through knee-deep snow to classes when the buses couldn't run. My last Wisconsin winter was 1938. In 1939 I was at Monmouth, and Jersey winters aren't exactly mild either. And I think in Texas I was as cold at 30 degrees above as I ever was at home at 20 below. And now I send pictures of myself swimming in the ocean on Christmas...

Jan 18, 1942 *(blackouts)*

I was pleasantly surprised at the way that phone call worked out the other night. The signal strength was equal to that on the local circuits here, which is understandable because it has to be amplified to be heard at all and naturally when they amplify it it's just as easy to amplify it enough as not enough. But I was surprised that there wasn't any fading or noise. When we call Borinquen Field, only 90 miles away, we usually have a pretty hard time hearing, but a conversation over more than 2000 miles works fine. It went by radio, you know, from San Juan to the States. . .

I saw Jack Benny in "Charlie's Aunt" the other night. I don't think I've laughed so hard at any show since "His Girl Friday" in San Antonio. When I left the theater, I found we had a blackout on. It was a practice one, and pretty poor. The papers the next day were full of how bad it was. Actually, it wasn't bad out in Santurce, where I was, except one store window that wasn't blacked out. It was locked up and no one could find a switch to turn the window lights off with. There was a mob outside who wanted to break the window to turn off the lights, but there were four Insular policemen standing there to prevent just that. I think they should have gone ahead and done it myself. People are just too indifferent about these blackouts and something like that would wake them up. Either that or one that wasn't practice. There was this big crowd, too, standing around by the lighted window, a perfect bomb target. I tried to tell the policemen to scatter the crowd and to explain why, but they didn't understand and I didn't argue because just then a reconnaissance car came by and I climbed in to ride down to the office. I have no inclination to drive my car during blackouts. I did out at Borinquen one night and it was no fun; out there the roads around the base are all dirt roads, there's construction on every one of them and in some there are big holes in the middle of the road -- I mean excavations, not just deep ruts -- and a lot of them aren't marked. . . In San Juan blackout driving isn't nearly so bad -- we have streets here, after a fashion, and all but official traffic is stopped during the blackout so it's really not too bad. But it's no fun regardless. I use official cars, with an Army chauffeur, at those times.

I've been swimming a few times lately, and the other day I played one set of tennis with Remington. The colonel has ordered that everyone will take one afternoon off a week to go get some exercise, but he doesn't do it himself and the rest of us are usually too busy too. When we do get away from the office we'd rather fly. Tomorrow I'm planning to go to Antigua to look over their radio set-up, and stop on the way back at St. Croix to look at theirs. There have been a lot of changes made at those places since I was there last.

Letters these days are hard to write. Everything you want to write about you can't. I don't know if any of my letters have been censored or not – officers sign their letters on the outside and sometimes the censor passes them without opening them. I suspect a censor unfamiliar with flying or the details of our operations here might cut some of the things I've said in this letter.

73

Aldro

Jan 29, 1942
Dear folks—
Another duty night -- .
3 hours later.

I always save my letter-writing for the nights when I'm duty officer on the theory that I'll be sitting in the office all night with nothing to do and might as well write letters; but I've never yet had a duty tour when I sat in the office with nothing to do but write letters. Sometimes things have gotten kind of exciting, and other times it's been just a lot of routine work that's piled up to keep me occupied. Tonight, we had a practice blackout, and I'm just getting over the work it caused me.

I've spent today putting up masts for some new receiving antennas; our station is revising its layout and going in for some new and fancy equipment. I've been thoroughly enjoying it; it brings me back to my ham days. That seems an awfully long time ago. Everything seems a long time ago. . .

Your letters aren't censored; only letters leaving here for the States are censored – well, of course letters leaving here for all points. But incoming mail isn't censored. I don't see the point in censoring mail going to the States. It isn't in the hands of any foreign government in transit.

2/7/42
Dear folks—

I've had a lot of catching up to do the last couple of days – I've been away on a trip, and things piled up. We spent several days in Panama, stopped at Dutch colonies in and near South America, Trinidad, and other places, a lot of which I'd never seen before.

I saw trees growing straight up and then going around a 90-degree bend to a windblown horizontal; I saw jungles like I'd never seen before, and swamps and mudflats where a forced landing would be pretty bad. I spent square nickels, rode on pontoon drawbridges, and saw cities with canals for streets. Nice trip.

In Venezuela or Colombia, or maybe both, are the Sierra Nevada de St. Marta. They're mountains, 19,000 feet high, ten degrees from the equator, snow-capped the year around. In one place the ground rises from sea level to 9,000 feet in a distance of about 9 miles – an average of 1000 feet per mile. From there to the top the average rise is more gradual, but quite rugged. There are cliffs a mile high.

All the mountains around here are cloud-covered a lot of the time; moisture-bearing winds condense their moisture to form clouds when they strike a mountain because the lifting of the air cools it and condenses out moisture. Above a certain altitude there's no more moisture to condense, so the top of this mountain is always in the clear. You can see it from 100 miles away. I flew through a pass at 9000 feet – it was quite turbulent.

Panama's hot and sticky; I wouldn't like to be stationed there. Facilities are good – Albrook Field has been there for a long time.

I commented to someone on the fact that the waitresses in the officers' mess spoke a lot like Virgin Islanders, and was surprised to be told that was what they were. When the Canal was built, laborers were imported from all over and the Virgin Islands were well represented. They speak a kind of English I've never heard anywhere else – soft, somewhat plaintive; their vowel sounds are very characteristic, but hard to describe.

2/18/42 *(Aruba & Curacao)*

Dear folks—

Things have been pretty busy here lately, as you can guess if you've been reading the papers. No threat yet to Puerto Rico, but enough work to keep us on our toes. I didn't mention having stopped at Aruba and Curacao on my trip, but with all the newspapers had to say, I guess I can describe the scenery without endangering anything. We spent only an hour at Aruba, and didn't go in to town at all. We spent one night in Willemstad, Curacao. (Pronounced cure -a-sah-o). Curacao is the seat of the government of all the Dutch West Indies -- including Aruba and Bonaire, nearby, and Saba, St Eustatius and St Martin, not far east of here. All the Dutch islands are outstandingly clean, and although their architecture is in general on the dingy side, like all the West Indies, its cleanliness is so impressive as to make the place look very pleasant.

Aruba and Curacao are important, of course, for their oil refineries, the oil from Venezuela being shipped in by tanker, refined, and then shipped out again.

The people there speak Dutch mostly; everyone speaks English too, and I was surprised to find a lot of Spanish spoken. . .

I climbed El Yunque the other day – drove to a point 2 miles from the top, & then walked. It's 3,500 feet above sea level, the trail is steep, and it's really a workout for anyone.

73

Aldro

3/4/42 (*"first enemy attack on American soil in the Atlantic"*)

This war gets crazier and crazier. The other night, a submarine shelled Mona Island for a while. It was probably in the paper in the States -- the papers here played it up as the "first enemy attack on American soil in the Atlantic". Mona is a small island halfway between Puerto Rico and Santa Domingo, in the middle of Mona Passage. It's a perfectly flat island, and nearly circular. All around it, the cliffs rise straight out of the water to about 200 feet, and the top of the island is flat. It looks like a big table top. I've flown low over it several times -- it's all covered with thick brush and there's hardly any soil on the island at all; just solid rock. It's full of wild goats and boars, and deer too, they say. The cliffs around the island are full of holes and caves where the surf has beaten away the rock, and the place is supposed to be full of forgotten pirate's loot -- it was famous at one time as a pirate's hangout. Kofresi, a famous pirate, and Captain Kidd both used it. At the west end of the island is a break in the cliffs and for a couple of miles there's a beach with enough land behind it so that a small airport has been built, and an NYA forestry camp. The airport is no good to anyone, it's just there. I've never landed there. Why the Germans should waste shells on a place like that, I can't imagine.

3/14/42

Col. House, our C.O., was made a brigadier-general today. We'd been expecting it a long time. He was passing out cigars about 10 inches long, and then he started walking around into all the offices, telling everyone to take the afternoon off. That "afternoon off" business is getting to be a joke. He issued an order long ago that we'd all take half a day off each week for exercise. As far as I know, no one's ever done it. Sometimes I've left the office an hour early to play tennis or go swimming, but as for taking the afternoon off -- well, it's one of those things you can't ever seem to get around to. So today the general circulated a note to all staff officers for us to initial and state whether we'd taken our half day off this week. All the answers were no. Tonight, the general said if that happens next week, he'll have us all tried by court-martial. Some fun! So now it's a military offense to spend too much time in the office!

3/19/42 *(promotion to Captain)*

Dear folks—

Biggest news of the moment is my promotion to Captain. I was at supper the other night when one of the waiters said I was wanted on the phone. It was my promotion. I came back to the table pretty elated, of course -- . Last night I was called away from the table for another phone call. When I came back and sat down, Remington wanted to know "What are you now, a major?"

You shouldn't worry about my being where the submarine warfare is going on. I can't think of any kind of fighting that involves less danger to airplanes. Besides, I hardly ever fly these days anyway. When the airplane's not away on a trip, I can't get away from the office. And when I could get away, someone else has the airplane. . .

It's too bad about Melvin Grinstead -- he's likely to spend about three months in the guardhouse. They may let him off easy, or they can make it tough on him. Since he said he wasn't going back, he's guilty of desertion, even though he may later have changed his mind. In peacetime, desertion for a period of less than 60 days is punishable by imprisonment for a year and a half -- and this isn't peacetime. If they call him just AWOL, the maximum peacetime penalty is 3 days in the guardhouse for each 24 hour unauthorized absence; and, again, this isn't peacetime. The least he can expect is to lose his pay for the time he was away (he'll lose that anyway) and he'll be reduced to $21 a month for a while.

I was swimming for a while this afternoon -- I just took an hour out from the office and took off. I went to the beach instead of the pool at the club -- I don't care so much for a pool. The main advantage of this pool is that it's only a few steps from my front door. The beach is about 1/2 mile away from the office. It was pretty nice over there today. There weren't many people over there and there was a heavy enough surf so that there was still a little force in it after it had broken over the reef into the beach area. There isn't any place in Puerto Rico that I know of where you can find real surf bathing -- all the beaches are protected by coral reefs. . .

There are always a lot of pelicans, it seems, over at this place. They just circle endlessly over one spot, a little to the east of our beach, about 10 or 12 of them. Every few minutes, one of them will wheel over on one wing, go into a dive straight for the water, fold his wings just before he hits the water, and go straight in. I like to watch them glide about a foot off the water, skimming up and down to clear the crests of the waves. One of the boys out at Borinquen tried something like that with a B-18 several months ago and came back with salt water dripping out of his tail. He was lucky.

3/26/42

If Aunt Johanna I had to pay 30 cents to mail that last letter of hers, the Post Office put one over on her. You can tell her that the airmail rate for first class mail to or from any member of U S armed forces anywhere in the world, is now 6 cents per half ounce.

73

Aldro

3/31/42

I got an Easter card from Aunt Helen. . . She told me about their trip through the South – said she didn't know there were so many negroes, shanties and moth-eaten mules in the world. She should see Puerto Rico ("Rich Harbor")!

4/8/42

I was glad to see you were able to get one of the good radios without too much trouble. It's a good-looking set -- I wish I could hear how it sounds. I think you'll enjoy the phonograph part too, although the automatic record changer is a gadget that takes a little learning how. The phonograph pickup uses a "permanent" needle, which is probably a sapphire. They last a long time, but they're easily chipped, and if that happens, they'll ruin all your records. . .

Saw a good show tonight -- "Dumbo." Have you seen it? Disney's feature-length cartoon about a baby elephant born with ears bigger than himself? It's awfully good.

4/21/42 *(the Haiti table)*

I bought a small table in Port-au-Prince which I intended to send you for Christmas, but I found it so handy I kept it. It has a reversible circular top about 18 - 20 inches in diameter, which sits on 4 folding legs about 18 inches high. It's Haitian mahogany, I think; has carved designs on each side of the reversible top, and as I remember it cost about a dollar or two. . .

I can understand Margaret's not caring to work in Chicago. I've never been to Minneapolis, but have never heard anything bad about it. I think the two most pleasant places I've been are Texas and the New York area -- Texas for the people, the country, and the climate, and New York for the availability of whatever you want. If she should go to San Antonio, she probably wouldn't stay single long. That town is called the mother-in-law of the Army. Puerto Rico would be about the nicest possible place to live if -- well, if. You can't beat the climate. If a hurricane were to wipe out all the towns on the island, the place as it might be rebuilt would really be the "tropical paradise" that the travel folders call it. That happened to Trujillo City in 1927, and it today is one of the nicest cities in the Caribbean.

5/5/42

My mailing address will be changed, on account of new regulations. You're supposed to address overseas mail now to numbered army post offices without giving geographical location, and my letters aren't supposed to indicate my location in any way. You're supposed to address my mail to Captain Aldro Lingard, HQ 6th Interceptor Command. APO #____, Postmaster, Port of Embarkation, New York City. Only there's no army post office serving our outfit, so you'll have to address me:

Captain Aldro Lingard, Army
HQ, 10th Naval District
San Juan, P.R.

I need a secretary just to keep my address straight. I just managed to get all my magazine subscriptions organized, and now I've got to change them again.

I flew a little while last night, but the weather wasn't too good and that nullified the purpose of the flight, so I landed after half an hour. It's the first night flying I've had since I went down to Trinidad to get General House, just after the war started. That was the time I took off about 2 hours before daylight, forgot my sunglasses because it was dark then, and took about two days to get over the beating my eyes took from the glare at 8000 feet, above cloud layers, for 8 hours. I like night flying -- you feel more alone, more away from things, in a world of your own.

Manzo and I played a little tennis this afternoon, but had to quit early. I don't play much tennis these days -- not nearly as much as I'd like to, but I get in a little now and then. They've just completed some courts just outside the BOQ; now I can come home from work, change in my room, walk 100 yards and play tennis, go swimming (pool), play ping-pong, or just relax in the sun. Except I never get home from work in time for those things.

5/17/42

Our speed on that flight last summer was about 140 to 150 mph -- normal speed for a B-18. It was a test flight for the purpose of checking the work that had been done on the automatic pilot -- you remember we flew quite a while with no one operating the controls. I don't know if I told you that the automatic pilot went out of commission again about two hours after we left Patterson on the way back. We had to fly all the rest of the trip manually. Those automatic pilots are hard to keep in shape. Incidentally, that airplane was junked about a month ago, after long and faithful service. Giannatti, about 2 years ago, had flown it into a thunderstorm, gotten into a spin, overstrained the wings and a few other parts getting out of the spin. They put new wings on and fixed her all up, but she's always handled a little differently from our other B-18's. So – about a month ago, they retired her.

My new address doesn't mean any move or impending move. It's just that it's been decided that it's undesirable to have soldiers' mail show the name of the organization and the location both.

5/29/42

I've been in Trinidad for a little over a week, in Port of Spain. Today I was to go back to San Juan, but the weather crossed me up. I'll get there tomorrow. . .

I'm adding continuously to my vocabulary of race-nicknames. I've told you how Mexicans are called spicks," and Puerto Ricans "gooks." Trinidadians are "jigs." It's short for "jigaboo." I think the term applies to the natives of all British West Indies. . .

You ought to see me drive on the left side of the road. After my stay here, and the time I spent in St. Lucia a couple of weeks ago, I'm an old hand at it. It's awfully confusing, though, trying to remember which of these places drive on the left and which on the right. All the British islands have left-hand traffic, and the Virgin Islands, but the Dutch possessions have right-hand traffic. And in Panama it's left-hand. Mucho confuso.

6/2/42 *(photography lecture)*

The last time I came back from a trip, I found I had a roommate. This time I found my office moved. Things certainly change. . .

I was glad to get the pictures you sent, but who took them, and who developed and printed them, and what kind of film did you use? I'm going to have to do some educating on such comparatively simple subjects as getting the subjects placed in the viewfinder, focusing, choice of backgrounds (mother's face is pressed flat against the south wall of the house on one picture), avoidance of distracting or poorly contrasting backgrounds, choice of film, etc. . . And don't place your subjects facing a bright sun -- A brilliantly lighted picture of a blinking, tearful subject is not very pleasing. Pictures taken in the shade on a bright, sunny day are actually better lighted than those in bright sunshine, and are much more comfortable.

Sorry to lecture -- thanks for the birthday presents – let me know Margaret's new address.

73

Aldro

6/18/42

Just got back from several places. Last week I went over to Antigua for a couple of days, and yesterday I got back from a short trip to Curacao & Trinidad. . .

You've got the sleeping Mexican bookends placed wrong. You've got them sitting side by side with books between, and they're supposed to be back to back with books between.

I see Elizabeth has taken up the Puerto Rican custom of wearing a flower in her hair. The girls around here seem to consider that the most important article of apparel they wear – you almost never see them without one. It goes well with dark hair.

7/2/42 *(Trujillo, Dominican dictator)*

Dear folks—

Yesterday I moved. The quarters I was in got crowded, so I'm now in the Cressione Apartments, an apartment hotel which the Army has taken over. It's a nice set-up – I've got a medium-size bedroom, another small room, and a bath. The small room is perfect for a darkroom – all I have to do is hang a dark curtain across the arch separating it from the bedroom. . .

Sunday noon I was asked if I wanted to take a trip to Cuba right after dinner, so I did. . . On the way back I stopped for a couple of hours at Port au Prince and . . . picked up a passenger who had to get to the States, and figured he'd have a better chance of getting there from here than from Haiti. He'd spent the last 6 months in Trujillo City, tutoring the Dominican dictator's son. I think I've told you something about Trujillo before – he's an ex-Marine officer who accumulated more and more power and finally set himself up as dictator over the Dominican Republic, named the capital after himself, & so on. His son, 13 years old, is a 4-star general in the Dominican Army – that's the equivalent to General Marshall, MacArthur, or Pershing in our army. Trujillo himself, the Generalissimo, wears 5 stars – we don't have any 5-star generals in our army.

We've got a Cub in our headquarters now – just got it today; I haven't flown it yet. It's a little bit of an airplane – one or two places, I don't know which, weight about 1500 pounds, horsepower probably 50 to 75, cruising speed something like 60 miles an hour. We got it so that we wouldn't have to use the big airplanes for one man to make short trips around the island, and I think it will be very valuable in that respect. Also, it can land in lots of places where the transport can't. Its usefulness compares to the transport as a bicycle compares to a cross-country bus. It will certainly be an experience flying it for the first time – I flew one a few times at Newport News, but it's been two years now since I've flown anything with less than 400 horsepower, and most of that time I've been flying 2000-horse stuff; now I'll climb aboard my 50-horse scooter and probably scare myself stiff.

Some of your incoming letters have been censored lately. Nothing's been deleted, except your reference to the Merrimac project in your last letter. The whole thing's the height of foolishness, I think; certainly they're not gaining anything by censoring the mail of their own officers, of whom our commissions all say that the President reposes the highest confidence in us, and in this particular case, I knew what the Merrimac project was anyway, and so does everyone who reads the papers.

73

Aldro

7/16/42 *(Miami – bridge in St. Louis)*

I've just spent four days in Miami on business; got back night before last, rather late. Miami's changed a lot. I suppose I'd have found the same thing anywhere in the States now -- at one time I knew all the bellboys at the Everglades Hotel in Miami, but now there was only one left that I knew. Childs' Restaurant, which used to be famous for its food and service, has apparently lost its key personnel; it's a disappointment to go there now. Walgreens too. Of course, there are still good places to go. I was pleasantly surprised to find a town that's not out to soak the army for every penny they can squeeze out -- well, maybe that's not quite a fair statement. Cities in the States have never been that way, but Puerto Rico and the other islands around here certainly are. Well --I walked by a theater one afternoon which had the price list hung over the ticket booth along with the sign "Men in Uniform 25 cents Anytime." That night I came back; they had the evening prices posted then, which were quite a bit higher, and the "Men in Uniform" sign wasn't there. But when I gave the ticket girl a dollar bill for a ticket, I got 75 cents change, and I never did pay over a quarter for a show in the four days I was there. In Puerto Rico, or San Juan, anyway, any theater you can get into for less than 55 cents is a hot, stuffy, smelly firetrap that you ought to be paid to stick your head in. Souvenir stores, too, advertise 25% off to men in uniform.

It's awfully hot there. I was almost glad to get back to Puerto Rico, where I can wear a shirt half a day and not have to wring the sweat out of it. Most of the important places in town are air-conditioned, though, which is a big help. I'd certainly like to transfer to the Air Corps headquarters there – they've got a whole floor of one of the biggest, most modern, air-conditioned office buildings, American stenographers, and all the telephone facilities that AT&T can give.

I shipped you a few mangoes. They're bigger than the ones that grow wild here, and may be better. I don't know, I didn't taste them.

I don't much care for the mangoes I've tasted, but I thought you'd probably be interested to know what they're like, even if you don't care for any more of them. . .

Your description of the Mississippi River boats pushing barges was interesting -- At Saint Louis, when we were away by ourselves and no one else was around, we used to buzz those boats sometimes. Fly low around them, looking them over. There were a lot of interesting things along that river -- Mark Twain was right. One thing I was always tempted to do was dive under the big Municipal Bridge and loop around it. I never did and I don't know anyone else who did, but it was everyone's ambition.

7/20/42 *(another promotion, and a good laugh)*

Dear folks—

Saturday, I got a letter from the Adjutant General saying I'm a major now. Some stuff! In peacetime you have to have 17 years' service before they make you a major. Air Corps promotions have been pretty fast lately -- much faster than in the ground branches, a fact which prompted some envious soul to hang a sign up above the bar in the Officers' Club at El Morro saying "Drinks will not be served to Air Force Lieutenant Colonels under 21 years of age unless accompanied by their parents."

7/30/42

Yesterday I went to Curacao and back – had an interesting trip. Just enough weather to relieve the monotony, but not enough to be troublesome. I'm in the middle of a cold, and I kind of dreaded that trip – I intended going all the way not over 1000 feet to avoid plugging my ears by altitude changes, but right off the bat I had to go to 7000 feet to get out of San Juan (weather), so I went the rest of the way up to 10,000 and then let down at 200 feet per minute at the other end – that takes 50 minutes, and you have to start letting down about 150 miles or more away from your destination. And I didn't have a bit of trouble with my ears, all the way down or back.

I want you all, and Margaret too, to send me complete dope on sizes of everything you wear; gloves, sweaters, shoes, stockings, and everything else. I was in Jamaica last week, and one of their specialties is doeskin gloves, which I could have gotten if I'd known the sizes. So if you'll send me all the dope, I'm going to make out a card, in tabular form, listing it all and carry it with me on trips. . .

I've lost my assistant – Willmore. They needed him worse somewhere else, and although I hated to do it, I had to let him go. He'll probably get a promotion out of it, although he doesn't know it yet.

I've found Consumers' Research very interesting all the time, whether I was considering buying the products described or not.

8/6/42 *(the other Haiti table)*

Dear folks—

I finally packed and shipped a small table I've been intending to send you -- it's an armchair table, or end table, or something, which I got in Port-au-Prince; it's mahogany with an inlaid top dash -- really a nice piece of workmanship, but some of this tropic-built furniture isn't too well seasoned and I don't know how well it will hold up in the States. I hope it doesn't warp, because it's really pretty nice -- the nicest I've seen around the Caribbean. . .There are some British pitchers in your package -- an "Old King Cole" milk pitcher with a cream pitcher to match. . .

I was in South America last week -- one of the boys bought a parakeet there and brought it back. It's a kind of small parrot, but it doesn't talk. Coming back in the airplane it got airsick and threw up. I've never heard of a bird getting airsick before.

8/20/42 *(sunrise above the clouds picture)*

Dear folks—

I was glad to hear you liked the stuff I sent you; did you find all the parts to the table? The three blocks that screw on the bottom to hold it up off the floor? . . . The pitchers I got in Jamaica. . .

I had a pretty nice trip. We left San Juan just before dawn Friday morning – I have some pretty good pictures of the sunrise, incidentally, from above the clouds, which I'll send you.

9/6/42 *(Angel Falls)*

I've enclosed a serial article on the AVG in China from Liberty; there are a few references to Dupouy. There's a clipping from the P.R. World-Journal of an Army airplane flying over an "unnamed waterfall somewhere in South America." The waterfall is Angel Falls, discovered a few years ago by an American pilot names Jimmy Angel who was prospecting for gold somewhere in the back country of British Guiana. It's the highest waterfall in the Western Hemisphere, but I've forgotten its exact height. Something over 1000 feet, I think.

9/20/42

I guess you know about our gas rationing. An "A" book is good for four gallons a month now. It's enough for me – I don't drive to work, and the only driving I do is to an occasional movie, and that's' not far. . .

Margaret says my slide rule is at home; I'd like you to send it to me. I quite often have occasion to use it now. Be sure to wrap it in at least two layers of strong corrugated paper, and I'd like it sent air express.

I'm sending you a Surinam half-guilder bill; it's worth about 25 cents American.

10/2/42

Seems I never stay home any more. I'm in Panama at the moment, and will probably go back to San Juan in a day or two . . .

I guess you might as well send my slide rule via boat mail, but make it first class. If you do that it may not be delayed more than a month or six weeks.

Has my transmitter gotten there yet? I turned it over to the Quartermaster several months ago for shipment, along with a box of clothes and things. When they do get there, you might as well open the clothes box – it's about 4 feet long and a foot square. All the other boxes are radio and miscellaneous stuff that can be left packed up. I think I'm going to have to send another shipment soon; I've still got too much stuff.

Yesterday I had the best malted milk I've had in over a year, except in the States. And it was made with powered milk. In P.R. we do have fresh milk, but it's pretty blue and the ice cream is nothing much more than sugar-flavored ice. Ice cream here isn't much good either, but in Trinidad, where they also have only powdered milk, you can get ice cream almost as good as that in the States. All of which seems to boil down to the fact that, except for drinking, good powdered milk is far superior to poor fresh milk. This is certainly a rotten area for anyone from a dairy state! What a Dutch Maid store would do for San Juan!

They don't wear ties here. No one does. Wonder when the powers that be will get wise and do away with them in P.R.

73

Aldro

10/15/42

Dear folks—

I just got back from Trinidad a little while ago, & I'm tired. In the last three weeks I've been on four trips, & have spent about four days in San Juan. Well, I think I'll stay home for a while now.

Margaret . . . seems to enjoy life in Minneapolis; it must be a pretty nice town, from what I've heard. I'm certainly getting tired of Puerto Rico.

You don't need to worry about my diet; I get plenty of butter, but no carrots, either raw or cooked. Tonight for supper we had a fresh pineapple salad. I weigh 185, and would weigh more if I didn't swim so much. Our doctor's desk is just across the room from mine, & he takes good care of me. . .

Beginning next month, I've arranged to have the War Department send you a check each month direct instead of my doing it. I'd like to know when you get the first one, just to make sure there hasn't been any slip-up.

73

Aldro

11/8/42

Have you seen Billy Gilbert in the movies? He's big, and fat, and usually plays comedy roles. He and a few other people from Hollywood put on a show here the other night, which I enjoyed. A minor faux pas on someone's part, though, was the inclusion in the troupe of a Cuban who sang in Spanish. He wasn't bad, but that kind of entertainment certainly isn't what the men who've been down here a couple of years or more would like sent to them <u>from the States</u>, of all places!

11/22/42 *(APO numbers)*

Dear folks—

I just got a letter from the bank saying the censor had told them they couldn't send anymore mail to "San Juan"; they had to have an A.P.O number. The Army Post Office number system was devised to conceal the location of our units from possible spies in the postal system, but succeeds only in hampering and confusing our own war effort, particularly when APO numbers change, as they often do. The silly part of it is that we use APO numbers now for everything. When I take off to go to Trinidad, my clearance doesn't say I'm going from San Juan to Waller Field. It says I'm going from APO #_____ to APO number_____. I can't even tell where I'm going without looking in the book.

11/28/42

I met an Army pilot who's a minister – that is, he was. They use him as chaplain in his spare time. First time I've known a minister in the army whose primary duty was anything but that of chaplain. This guy's a combat pilot – his job is to bomb submarines.

Someone's radio is playing "I'm dreaming of a White Christmas." That's certainly a beautiful song. You know, I've had two attacks of particularly acute homesickness in these two years – one when Wayne sent me those pictures he'd taken of all of you with the new radio, and one when Alan sent me some snow pictures he'd taken in Idaho – some of the most beautiful snow pictures I've ever seen. It's nearly three years, you know, since I've seen snow.

12/8/42

Did I tell you I have a new office? Air-conditioned, quiet, plenty of room (for the time being), & everything. Moved into it last week, & after enjoying it for one hour, left on a 2-day trip, stopping at Jamaica, among other things. I slept under a blanket in Jamaica, and today I got a blanket from the squadron for my own bed. I haven't had a blanket for nearly a year – didn't need it. I don't expect to need it now, but I'll have it in case I do.

What gave you the idea we don't get enough to eat down here? Did I? It's not always what I want, but it's always something -- plenty. I weigh 185 pounds, and should weigh less. And I certainly don't fly hungry -- on any flights over three hours, we usually carry sandwiches, Coca-Cola, fruit, and spend more time eating than flying. Didn't I tell you about the time, coming back from Cuba, when I had to stop eating tangerines because I hit some instrument weather, so Colonel McKinney stood behind me and peeled them and fed them to me?

Food is a problem for the natives in some of the outlying towns, and even in San Juan, I've heard, civilians have trouble getting butter & eggs. . .

All the music around here is of two kinds: either it's loud, repetitious, blaring, clattering music of the type that's mistakenly referred to in the States as "romantic Latin rhythm," or it's some guy singing in Spanish while he strums the same succession of guitar-chords over and over again. You see guitars all over here. You drive along a country road on a Sunday afternoon (or you did before they rationed gas), and see any number of people walking along, alone, singing and playing a guitar.

12/25/42 *(first mention of Mom)*
Dear folks –

Merry Christmas! I walked into the office this morning and the adjutant (Colonel Pheberg) wished me a happy new year. I asked him why and he said, "Well, it doesn't seem like Christmas."

Thanks very much for all the presents; I was surprised at the variety of things you were able to think of. . .

Did I tell you I've been reading a book "Admiral of the Ocean Sea" about Columbus? I find it particularly interesting because I've flown over almost all the places where he stopped. I was surprised to find that the idea of a spherical Earth was apparently pretty widely accepted in his time; the principles of celestial navigation, which are based on a spherical earth, were apparently quite well known, and the terms "latitude" and "longitude" had been in use for a long time. With the instruments he had, Columbus couldn't navigate at all accurately, but at least he knew what he was trying to do. . .

I've been trying some portrait photography again. My last attempt at that was back at Monmouth and was a miserable flop. How do you like the enclosed picture that I took of one of our stenographers? (Not mine.) Her name is Conchita Hernandez -- kind of a Spanish equivalent for Mary Smith. We'd been working almost half an hour in my room, taking different poses and trying different lighting schemes -- they all turned out later to be not much good -- and then as I started to put the camera & things away, she flopped into my big easy chair and leaned over the back of it to watch, so I took this shot that way. I like it pretty well. The right arm and hand look a little too big, though, being closer to the camera than her face. I may try that shot over again.

1/18/43 (*He took a stenographer up in a plane. Was it Mom?)*

Your pictures came through, and I was glad to get them. The part of your letter telling how deep was the snow in the pictures was censored, though. One of the fellows here told me a letter from his folks had had a few words about Lake Ontario freezing over cut out.

I think I told you about the six hours a day of English on the local broadcast stations, including transcription broadcasts of all the big programs like Jack Benny. It's pretty nice; makes sitting at home evenings to listen to the radio worthwhile. Right now I'm listening to a program of organ music by one of the officers in one of our support units, and enjoying it.

Yesterday I took one of our stenographers riding in a Cub -- first time I've ever done that. It was a lot of fun -- we were flying along and all of a sudden we had a couple of wing men -- two other Cubs had tacked on and were flying formation with us. One was a Navy officer and the other one I didn't know. The ship I had was a Cub Coupe, one of those with side-by-side seating. It's more like an automobile inside than anything I've ever flown. It was the first time I had flown that kind of ship. . .

I've just seen a new pilot's insignia. In the Army we have three kinds of pilots of different amounts of experience: Pilot (like me), Senior Pilot and Command Pilot. A Senior Pilot has to have five years' flying service, and wears a star above the shield on his wings. A Command Pilot has ten years' flying service, and wears a wreath around the star above his wings. Anyway, someone in the headquarters here designed a new insignia: in the new Pilot's insignia the shield has been replaced by a swivel chair. The Senior Pilot's insignia has a typewriter above the swivel chair, and a wreath of official-looking papers surrounds the typewriter to make the Command Pilot's wings. Well, that insignia certainly fits those of us here in San Juan.

73

Aldro

1/28/43

I got back from Camaguey yesterday and found a wedding announcement from Alan. Shortly after the chain letter started, someone, I think it was Alan, started wondering in what order the members would get married. Most of us thought Alan, or maybe Wayne, would be first. I think we all thought Carl would be last or next to last, and actually, just about everyone's whole list turned out to be upside down. Well – that leaves only me; I imagine the chain letter will have made several rounds before the boys get tired of asking me – "Well?". There was also an agreement that each would write an essay, when the time came, on "Why I Got Married". Carl's was a masterpiece, like everything he writes. Paul hasn't gotten around to his yet; Wayne's was straightforward & interesting, and Alan's is now due.

Was looking through a file of Airplane Commander's reports the other day -- the report a pilot fills out when he comes in from a tactical mission telling what he did, where he went, what weather he had, and any unusual or important occurrences. Most of the reports were pretty routine, of course, but occasionally someone would cut loose. One started to quote poetry –

"Out of the night that was black as pitch –
Out of the smoke and the din and the roar –
The left engine coughed and we came home."

Another, engaged to a girl he hasn't seen for I don't know how long, reported, under Weather –"One helluva gorgeous moon out there."

73

Aldro

2/11/43 *(definition of will power)*

I guess I must have eaten close to a pound of salted peanuts in the Officers' Club at Zandery Field last night, talking to a bunch of guys. You know the definition of will power – "the ability to eat one salted peanut". We can't get them in San Juan.

73

Aldro

2/18/43 *(first water landing)*

Today I made my first water landing – someone got hold of a pair of pontoons and put them on our Cub, and I was out flying that around this afternoon. Seaplanes are a lot of bother though – they don't handle well in the water, they have to be washed off after every flight, and launching them and beaching them is complicated and takes a lot of men, even for a Cub. Besides that they fly slower than landplanes, and are more sluggish in the air as a result.

3/9/43 *(nurse Mom)*

Dear folks –

I'm writing this in bed: I've had a cold the last couple of days, and although it hasn't bothered me much, this afternoon up at the office I could feel myself getting hoarser each time I answered the phone, so I knocked off about 2:30 & came home & went to bed. . .

I was sitting up in bed working on some stuff I'd brought home from the office when, about 5:30, there was a knock on the door and in came Conchita, -- the girl whose picture I took looking over the back of my chair. I'd told her why I was leaving the office early, and here she was all loaded down with packages, a small kettle, cup and saucer, and a few minutes later a boy showed up with an electric hot plate she'd borrowed somewhere. I haven't had such a going over since long before I left home. Sal Hepatica (ugh!), warm Salymentol (like mentholatum) rubbed on my chest, hot lemonade (and lemons are hard to get) – you'd have thought I was sick. She stayed here about an hour – missed her own supper probably, because she lives pretty far out in the sticks. She lives with her sister; their mother, who lived with them, died a couple of months ago.

4/1/43

I've been getting in a little more swimming lately than I'd been getting for a while; I usually get over to the beach a couple of times a week now. Three people have told me independently in the last two days that I look like I'm losing weight, but if I am it's not much; I haven't weighed myself to find out. Several months ago I weighed about 180, which is what I haven't deviated from very much in several years.

4/22/43 *(in the hospital)*

At the moment I'm trying with what I know won't be any success at all to get out of the hospital. Sunday I felt a cold coming on; Monday I felt worse, although the cold itself hadn't shown up yet, and had a slight fever. Tuesday was the same. So yesterday, having listened for two days to the General and our Flight Surgeon tell me I ought to be in the hospital, I came on up here. Well, my symptoms are all gone now; I feel fine, but they won't let me go. Not that I really expected them to. No one ever gets out of an Army hospital in very much less than a week. It irks me though; there are two reasons why they can't do any good by keeping me here: one is that I feel fine, so they can't make me feel better. The other is that since my symptoms are all gone, there's nothing left to tell them what was wrong with me. I think myself it was just a little flu.

Did I tell you I was in Panama last week? Spent several days there, & was really glad to get back here where the weather is bearable. I'm glad I'm not stationed there. I didn't see Carl – he was in the States; he'd finally wrangled some leave to go home and meet his several-months-old daughter.

You've heard all kinds of stories about the French penal colony at Devil's Island. I saw it on that trip I mentioned. It doesn't look nearly as bad as the mainland.

4/28/43

Dear folks—

They still won't let me out of the hospital. However, they think they know what I've got now, so I don't mind so much. It's MONONUCLEOSIS, an unusual condition of the white corpuscles which I don't understand, although the doctor showed me what his book said about it. The symptoms & effects are pretty much like flu, & it usually lasts 2 to 3 weeks. I've had it nearly 2 weeks now, and since mine is a mild case I think I ought to be out of here pretty soon now.

I mailed your Mother's Day present yesterday – or rather, I had it mailed. Conchita wrapped it & took care of getting it off to Miami. She helped me pick it out, too.

5/6/43

Dear folks

Well, the people who claim to know better than I do still say I'm sick, and won't let me out of the hospital. My temperature's been normal for nearly a week, but my white blood count still shows too few polymorphonuclear cells, so I'm still here. I can get out for a while – Sunday I went to a show, & last night to another.

Did you hear the story of the patient who was due to be discharged from the hospital and didn't want to go because he was in love with his nurse? The morning he was to leave, the nurse took his temperature as usual, and when she wasn't looking he took the thermometer & rubbed it on his pajama sleeve a few times so the friction would raise the reading. Then he put the thermometer back in his mouth, the nurse came back and wrote the reading in the book and went away without saying anything. Pretty soon she came back with an orderly who gave the guy his clothes & started getting him ready to leave, whereupon he expostulated "What are you doing, throwing out a sick man? Didn't you see the temperature I had? I can't leave here now!" And she told him "Oh yes you can. Your temperature was 120. You're dead."...

5/17/43

I may be home a little sooner than I'd planned; I'm still in the hospital, probably will be for a few more days – maybe a couple of weeks – and then the doctors want me to take some sick leave right away instead of the regular leave I was going to take in a month or so. My blood count is still holding me in here, but that's all that's holding me; it's just a question of waiting until it gets back somewhere near normal, which it's doing slowly. I come and go here pretty much as I like – go to the office for a while, go to a show – there's no real reason why I should stay here, except they still think I'm sick...

I lost my boss. General House has been transferred back to the States, & will go to a combat zone. I was sorry to see him go; I don't think there's an officer on his staff who wouldn't have liked to go with him.

6/29/43 *(on his return to Puerto Rico after extended sick leave at home)*

Dear folks –

Got in this afternoon, after a trip partly pleasant, partly hectic, and definitely varied...

I found a warning order when I got here telling me I'd be moved soon – from its tone, I think Gen. H is pulling strings thru the War Department to get me over there with him. Col. Serig will go too. We don't know when, but it will be soon...

It will be a big change, & a rather welcome one, although I'll miss a lot of things I'll have to leave behind me.

Wedding Announcement

7/6/43
Dear folks –

I'm sorry not to have written sooner, but I've – we've – been pretty busy the past few days. We still are.

Friday night I tried to call you to tell you Conchita & I were considering getting married & to ask you both what you thought. The call didn't get thru until Saturday morning, and even then I didn't get to talk to you both as I'd wanted to. Anyway, by that time we'd made up our minds, and since I was to leave very shortly, it seemed that every day we waited was a day wasted so we didn't wait. It was Saturday & the marriage license bureau was closed, but I got hold of the man that runs it. We needed physical examinations, including blood tests which usually take several days but which we got done in two hours. We needed written permission from my Commanding General and from our next higher headquarters, which can take as much as a week to simmer through but which we fixed up in short order. Neither of us wanted a big wedding, so there were few problems and no regrets in that connection; we asked a couple of friends to be witnesses, got our own chaplain lined up, and didn't tell anyone else until afterwards. Less than 24 hours after we'd made our decision we were married – 9:00 PM Saturday night, July 3.

That's when the fun began! I thought we'd done a pretty good day's work, but one thing we hadn't gotten around to was arranging for a place to stay – the Crescioni is a bachelor establishment so we couldn't stay here. It hadn't worried us because you can always get hotel rooms; however, this was Saturday night on July 4 weekend, and all hotels were jammed. Things were beginning to look awkward when we finally managed to grab off the last available room in a third-rate hotel two blocks from the Crescioni. The next day we were able to get a nice room in San Juan's newest hotel, & the following day we got permission to stay in the Crescioni, since it would be only a few days. It looks now as though I may get to stay here a couple of weeks yet.

In not telling you sooner, I really wasn't keeping anything from you. I'd never intended to get married, ever. And although Conchita & I'd been seeing quite a bit of each other since last December (she'd been working in our office then for about 4-5 months), we hadn't considered getting married at all. They say that absence makes the heart grow fonder, though, and I soon realized that the way I felt when I got back here from home wasn't very consistent with my earlier intentions, so --- It's a poor time, of course, it's too late to do it sooner, and later is too far away and too uncertain. We have no regrets; we're very happy.

At present Conchita is on leave from her job; after I leave, she'll go back to it. She may go to the States later. There are some things in the safe deposit box I'd like you to send Conchita to keep – my bank books, and my government insurance policy, which I'm going to change making her the beneficiary. The other two policies I don't intend to change, and you may as well keep them. On second thought, my First National bank book is the only one of those that I want; the Union Trust & Mt. Horeb books may as well stay in Madison.

I left the picture frames behind because they were too bulky to take or send. They're OK there.

I'm enclosing a letter from Conchita.

73

Aldro

Dear Mother and Dad: *(Enclosed letter from Mom)*

Just two or three lines to let you know that Al is feeling fine and I am trying to take the best care of him. I want you to know that the fact that Al has selected a wife does not mean that you have lost a son, it means that you have another daughter and I hope to occupy a little corner in your hearts as such. I'll try to be a sweet and worthy wife and offer him all the happiness he deserves.

Now that Al is going away and I'll feel so alone, I'll like to receive words from you often. I may be able to go to Madison to see you in the not too distant future. In the meantime I'll be thinking and loving you.

Please convey our love to the girls, aunts and uncles.

Lovingly,

Conchita

7/11/43. *(Handwritten at the top: "About wedding. Keep as it is for book." Grandma must have intended to write a book. Portions in italics below were in Mom's handwriting.)*

Dear folks –

It was Thursday nite we talked to you, and only now – Sunday – that we find time to write. I'm really not busy at the office – just all kinds of little things that keep us both tied up. Conchita was awfully glad to talk to you – we called from a special international phone booth where they have provision for several people to listen in, so we both heard everything.

Tuesday, we went to St. Thomas and spent the night at Bluebeard's Castle. Although the building is stone, there are a lot of wooden shutters & things that made noises in the wind. It was a rather stormy nite, & with shutters rattling, distant thunder, a lonely setting, and the name "Bluebeard's Castle," Conchita was pretty scared most of the night. We enjoyed it though; the next day we bought souvenirs, went swimming, & then flew back here. I got wedgewood pitchers for Dolores Pentzborn & Rosie, and one which Conchita's going to keep for us. Those were the last three they had, incidentally. I've got some other things Conchita's going to keep – my easy chair, my radio, a little Haitian table, my book-case, and some of my books. Most of my books, and my magazine files I'll send home, and I'm going to change my magazine subscriptions so that most of them will come directly home, since my technical magazines are of value mostly for filing.

We had some wedding pictures taken, of which we'll send you prints. Also, we've taken a few informal shots of ourselves, some of which we're going to develop tonite. This afternoon we're going swimming. We are also enclosing clippings announcing our wedding. We both think it is terrible the way they announced it. Conchita's folks were going to send the announcement to the press but someone from the "World Journal" sent it out without complete information about us. Well, too late to do anything about it.

I (Conchita) was all excited and thrilled when I talked to you Thursday night. To say the least, that is the most exciting conversation I have had and I hope my husband does not get jealous if I say that is the one I've enjoyed most (including the conversation I had with him while on leave).

I want to assure you that Al was not keeping any secrets from you. We did not think of getting married but while he was gone I guess we found out we were in love with each other. You have no idea how very much I missed him and how very hard I prayed he would come back soon. Now that we are married I am sure I love him more than anybody else could love him and I'll do anything under the sun to make him have complete happiness. He is wonderful and I pray to God nothing makes him change and also pray he be sent to us safely and be kept from all danger. I hope my prayers are heard by our dear Lord. He will come back to us with honor and glory. He is a good soldier and his General knows it, so we should be proud of him. It is his fate that he goes to face the enemy and we must not question that. God knows best.

(back to Dad's handwriting)

I'm going to have some more copies made of the pictures I had taken in Madison, & I'll include in the order a picture to be delivered direct to the church. Right now Conchita's got the proofs – rather, she left them at her sister's house in Rio Piedras. Incidentally, you and she agreed on which pictures were best; as I did with you, I showed her the five proofs & asked her which she preferred. She picked the same one I'd already picked, so then I unwrapped the copy I had of it and gave it to her (same size and kind as the big one on the book-case). Of the three in which I wore a blouse, she also preferred the same one you did.

I don't leave here until the 24th; I have a few days leave for a honeymoon, but we're spending most of it here in San Juan.

We haven't written the girls yet, but we will and will send them copies of our wedding pictures. My husband is ringing the bell and says it's time for swimming so we will say until next.

With all our love,
Al & Conchita

P.S. I don't know where Al put the clippings I had for you, so we will send them with our next letter as we want this to leave early tomorrow.

Where **I** put them! I didn't have them! I think the maid threw them away. We'll try to get some more.

Al

7/21/43

Dear folks. –

The last 3 months have been pretty eventful for me.

1. I got sick
2. I went home
3. I got married
4. I'm going to Africa
5. I got promoted
6. I got an Air Medal.

I'll be leaving here Saturday, but I've still got a lot of packing to do. There'll be some string rugs in the boxes I send home – they're not presents this time; Conchita & I selected them for ourselves. However, if you want more of them, or of anything else, she'll get them for you.

My promotion came thru last Sunday. Conchita was more excited about it than I've ever been about anything. We were to have a picnic that day with some of her brothers & sisters – went out in the country & ate roast pig and went horseback riding. They roasted a whole young pig over an open charcoal fire; it was awfully good. Also we had some native coconut in the only form I've ever tasted that I like – they take out the meat, shred it, and cook it in a sweet sauce & eat it for dessert. It's very good. . .

The Air Medal is awarded for "meritorious achievement while participating in aerial flights." There used to be only one such decoration, the Distinguished Flying Cross, which was awarded for things like unusual skill, bravery, or cool headedness in meeting such emergencies as forced landings, etc. or, since the war started, for definitely sinking an enemy submarine (you'd be surprised how much agony that word "definitely" makes). But they needed a lower-ranking decoration for less meritorious achievements such as flying long hours of anti-submarine flights over a long period of time when, through no fault of his own, a man doesn't get to sink a sub because there aren't any of them. He may have done a good job of flying, and most likely he's been out over the ocean for a lot of hours in bad weather or on dark nights. It's hard work and deserves recognition. It so happens, though, that they wrote the regulations so that just about any kind of flying over water counts toward the Air Medal, with the result that, for flying a nice new safe comfortable airplane in nice weather between various point in the Caribbean whenever I felt like it, I've got a medal that was intended to be given for an entirely different type of flying. I haven't been in any danger, or under any pressure to take off in bad weather or at night, and about all I've done is let the automatic pilot fly the airplane while I've listened to the radio. So, I can't feel very proud of my decoration.

Conchita's got a miserable cold – oh oh here she comes. Now she's on my lap. This is the only comfortable chair we've got. Did you ever play rummy with your opponent sitting on your lap? I did the other night. . .

Later

I'm curious as to the relatives' reactions about my getting married – especially Aunt Jo & Drevdahls.

Conchita's already gone to bed (6:40 pm) – and doesn't feel like writing. Her cold has her feeling pretty bad. Incidentally, when you write to her your letters will need 10 cents postage for each ½ ounce; letters to me will still go for 6 cents per ½ ounce. My letters will be irregular for a while. Incidentally, V-mail will probably go faster than regular mail.

73

Aldro

(in Mom's handwriting)

I feel terrible but not too bad to send you our love and lots of kisses. We'll write a long letter tomorrow. Sweet dreams.

Conchita.

CHAPTER 6

Letters from North Africa

July 1943 – January 1944

Nowhere in these letters does Dad say where he was stationed in North Africa; if he had, it would certainly have been censored. For some reason, I grew up thinking he was in Morocco, but this passage from a summary of his career that he wrote at Grandma's request in 1956 suggests that he was based in Tunisia. Other comments had me leaning toward one or the other. See what you think. The October 29 and May 1 letters may be hints.

"When I first went to Africa I stopped en route at Dakar, French West Africa; Tindouf, an isolated "Beau Geste" type station near the border between French West Africa and Algeria; Marrakesh, just across the Atlas Mountains in French Morocco; Oran, Algeria; Algiers for a few days; Constantine, Algeria, for a few days; then finally La Marsa, a resort town a few miles north of Tunis on the beach, near ancient Carthage."

7/25/43
Dear folks –

I left San Juan yesterday, & am now well on my way where I'm going. Tonight will probably be the last I'll spend in the Western hemisphere for some time. . .

Being married to a girl certainly changes the way you feel about her. I missed Conchita an awful lot while I was home, but that was nothing compared to the way I miss her now. Of course, I miss you all too, but I've gotten used to that. Well, we had three weeks together – exactly 3 weeks – during which time we were never separated at night and very seldom in the daytime. She's gone back to live with her married sister now, and goes back to work tomorrow.

Monday morning –

I'm in Brazil – ran into a friend of mine last night who's stationed here. He's going to show me the town this morning.

8/2/43

Dear folks –

Tonight I'm in a French hotel in North Africa, writing by the light of a single bulb in the middle of the room (I'm in the middle of the room directly under it) with all my clothes off so I can be cool. I haven't got to work yet, but at least I know now what's what – temporarily at least – and will get to work very soon. . .

I was cautioned that the Moslems in North Africa don't like to be photographed – they believe that a camera is an evil eye, and that a person who has a photograph of them can do them harm by putting a curse on the photograph. Something like that, anyway. So I was very careful. Then I made the rounds of one town with an officer who'd been there long enough to know, and he told me they didn't mind at all. So I snapped pictures right and left, and sure enough, they didn't mind at all. All except one bearded, dirty, sheet-wrapped old man whose picture I didn't even try to take. He happened to be sitting where I pointed my exposure meter to get a general light reading. He jumped to his feet and started jabbering, and finally walked away, still muttering and throwing me dirty looks. He was probably cussing a blue streak in Arabic. . .

I got what should be some good pictures of an Arab water peddler, and of a snake charmer with two carpet vipers and an African cobra.

I miss Conchita, of course – more than I'd thought I ever could miss anyone."

8/12/43

I was in Italy the other day. I flew over the crater of Mount Vesuvius – it has three craters really. There's the old one that destroyed Pompeii (I guess), a newer one about the same size, & perched up on the lip of the newer one is a small one from which steam pours continuously. I think they use that steam to generate power. . . Next time I go over there I'm going to take my camera along; I understand the censors don't like aerial shots, but I can't see any harm in this one & will take a chance on it.

Starting below, Dad switched from dating month/day/year, like we do in the States, to day/month/year, as they do in France and presumably North Africa.

5/9/43

There are several languages spoken here – French, Arabic, a mixture of the two, and a very little English. I have a French book, but haven't studied it much. I can get by – about all the French I need is enough to ask for "d'leau" (or is it "du l'eau") in the mess instead of the sour lemonade we have at every meal, and to ask for it "tout suite" the next half dozen times the waitress breezes by. And I don't need any Arabic at all.

Have I told you about the Arabs? They wear long, voluminous robes, & the women wear hoods & veils so all you can see is their eyes. However, when I lived down on the beach (I moved a few blocks about a week ago) some Arab women used to come down every morning to the beach in front of my villa, take off their veils, robes, & everything but a scanty underskirt, & go swimming. The fact that the beach was by no means deserted didn't bother them a bit. . .

Besides Col. Serig, Krisberg is here too; he turned up the other day.

I got back yesterday from an overnite trip to Algiers, & the airplane I used was so fast it took – well, if I'd tried to fly that Cub we went up in half as fast as the speed I made on this trip I'd have pulled its wings off.

Did I tell you I was in Sicily a few days ago & saw Bob Hope put on a show there?

73
Aldro

10/9/43

Today I got your letter postmarked Aug 2. It took 39 days to get here. . .

I was out on a short trip this afternoon. These ships we've got around here are really fun to fly – last week I went to Algiers at an average speed well over 300 miles per hour, and I was flying the same ship today.

I've been wondering what the reaction around home was to Italy's surrender. . .

Did I tell you I've been driving a German jeep (Volkswagen) around lately? It's a pile of junk, but does surprisingly well. Reminds me of my old Chevrolet.

13/9/43 *(description of Mom)*

You wanted a description of Conchita – she's 5 feet 2 inches tall, weighs 112 pounds, has black hair, complexion about like mine, will be 26 years old December 8. You mentioned her teeth – she brushes them about half a dozen times a day. She wears size 4 ½ shoes, 8 or 8 ½ stockings, 14 or 15 dresses, 8 gloves, 36 pajamas or nightgown, 34 sweater or slacks (however, as long as she stays in PR she can't wear slacks); has a 34 ½" bust, 26" waist, 36" hips, -- I took all those measurements before I left in case I should run across things to get around here.

16/9/43.

I have no pictures of Conchita & the car – they must be some she's had taken. I imagine she's sent me some. It's a 1940 La Salle that I got last December from Col. Harvey, my old squadron commander from Langley. It's the smoothest-riding and easiest-driving car I've ever driven, has a good radio, seat covers, fair tires (better than average) and a new paint job that I had put on while I was home. I hated to have to leave it. It has a heater and windshield defroster too, which I never used, of course, and a lighter & ash trays – it's got everything. . .

Jack Benny & some other people are putting on a show near here tonight. Col. Serig & I are going in the Volkswagen. That thing is quite a vehicle – reminds me of a Model T in a lot of ways. Mostly the way it rattles and bounces. It's certainly different from my La Salle.

27/9/43

Conchita told me you'd suggested her coming to Madison, but that she didn't feel like going until I come back. I most definitely don't want to have to stop in San Juan to pick her up on my way back – it might not even be possible. I think the thing can best be settled directly between you & her . . . I'm not likely to be home before spring.

I've been out flying today – tried out another airplane of a type I'd never flown before. You should have seen my first landing! It wasn't bad enough to be dangerous, but I'll never brag about it! My other landings were OK. I don't do much flying these days, which makes me enjoy so much more the flying I do get to do. . .

I'll be interested in your comments, when I get them, on the Italian operation; how do the people in the States feel about that? I can't tell you, of course, what we think of it, other than to refer you to the newspapers.

23/10/43

It seems funny to see everything getting green around here, just as if it were spring instead of fall. All summer it never rains, & you see nothing but sand. In October the rainy season starts & the trees get green, grass comes up, & it looks like spring. . .

I was out flying a B-17 the other day & had an engine go bad on me, so when I came in I had to land with 3 engines. It's no trick in a B-17 & not dangerous at all; it's interesting only because it's the first time I've ever had engine trouble in flight. At this rate I can expect to wait 4 more years before it ever happens again. Actually, if the thing that went wrong had happened in a single-engine ship I'd have kept the engine running, because the trouble wasn't serious. In this case, though, I had three other engines so I could easily afford to shut down the one that was cutting up. Rather an enjoyable day, all in all. .

Got a letter from Margaret the other day; it took 34 days to get here. . . Airmail from you averages about 25 days; from Conchita about 35 days. My letters to her take about 20 days.

73

Aldro

25/10/43

You don't need to worry about cholera – I've had shots against cholera, tetanus, typhus, typhoid, yellow fever, and bubonic plague. And smallpox. That didn't keep me from getting a touch of jaundice a little while ago, though. I didn't have it badly, & I'm all over it now.

Col. Serig & I visited *(word cut out of letter, presumably by censors)* yesterday; for the Moslems, five pilgrimages to *(word cut out of letter)* are equivalent to one pilgrimage to Mecca. We saw the inside of one of the older mosques – it was beautiful. It had two domes, both of which were finished on the inside with plaster filigree work all over. Under one of the domes was the tomb of a friend of Mohammed's; all over the floor in that room were prayer rugs – small scatter rugs, heavy, woven in designs from the Koran, donated to the mosque by pilgrims to the holy city. For a holy city, though, it was certainly dirty.

73

Aldro

29/10/43 *(Kairouan is in Tunisia – this time it wasn't cut out of the letter by a censor)*

I haven't done much travelling lately – so no, I haven't yet been in Italy & I've only been to Sicily once. I told you about my trip the other day to Kairouan, one of the Moslem secondary holy cities.

I can't tell you where I am, but Conchita can give you a good idea. You know, I read a story about a soldier who spelled out the name of the place where he was by signing a different middle initial on successive letters home. The letters arrived in the wrong order though, & he got back a very puzzled question: "We can't find this place "NUTSI" on the map anywhere.

2/11/43

Don't worry about my being in any danger – this is the safest place I've been in yet. Disgustingly so.

21/11/43

My blood count is OK – I had it checked when I had jaundice, which, incidentally, one of my officers now has. It's pretty prevalent. Not serious, of course, but in order to get over it you have to knock off work for a while. I had only a mild case, but was in the hospital two weeks, during which time I felt fine, read a lot of books, was well looked after, saw a good show nearly every night & enjoyed life generally.

The mountains are getting snow caps; they all start at about the same elevation, extending upward, so that from a distance, seen against a blue sky, they seem to be sticking up from beyond the horizon. It's pretty. I flew over several snow-covered peaks the other day; in the cockpit I was warm & comfortable in ordinary summer flying clothes.

23/11

I've been getting in a little more flying than usual the last few days, but expect the law of averages to catch up with me & keep me on the ground for a while now. . .

I subscribed to the Readers' Digest in August, & my first copy (October) finally came yesterday. Now that I've finally got a fair supply of things to read, I no longer have so much time to read them.

73 -

Aldro

30/11/43

Last week I ran across one of the gang I'd known at Monmouth in Algiers. He's a Lt. Col now – had been thru the Sicilian campaign with Gen. Patton. I hadn't talked to any ground officers about any of the operations around here – I found his experiences and viewpoints pretty interesting.

I was looking at a round brass plate in Tunis the other day with etching all over it and Arabic writing inlaid in copper and silver. . . I'd seen so many things with Arabic written all over them I was curious – I know most of them were quotations from the Koran, & I think I've told you that the designs on most of the rugs & tapestries are also supposed to come from the Koran. So I asked the clerk what the inscription meant. He said it means "Souvenir of Tunis".

The box of apples I got in Naples is almost gone; looks like I'll have to make another trip. . .

I got two packages from Conchita the other day. One was an electric razor I'd asked her to get me, & the other was a big assortment of stuff . . .

I asked her to get me the electric razor because I was getting tired of shaving in cold water. I've never cared for an electric razor because it doesn't shave close & it takes a long time to get a barely presentable shave with it, but close or not it's certainly an improvement over cold-water shaving, and the day the razor came, it happened there hadn't been any water at all so I hadn't shaved that morning. That doesn't happen here often, but sometimes. It seems that wherever I've been the supply of electricity for shaving was more dependable than the supply of water for shaving. Not that I mean to brag about the electric supply in the building I'm living in now – it's a 10-room villa, with servants' quarters besides, & the whole works is supplied thru a 5-ampere fuse. Five amps! That's enough to run one electric iron & no more! I can't understand these French. They build big, beautiful villas to live in, with tile floors & masonry walls & nice flower gardens in the enclosure outside the house. They put in stained glass skylights and fancy electrical fixtures, and then they put in a 5-amp fuse? To them a 25-watt bulb is huge – they light (so they call it) a whole room with one 15-watt bulb. An electrical candle, practically. . . the picture's just not consistent; presumably they use small bulbs to save on electric bills; and yet this villa, if built in the States, would cost, I imagine, about $15,000.

14/12/43

I've never been thru the Casbah in Algiers, but I went thru the one in Tunis not so long ago. The streets (called "souks") are roofed over, & the houses are so close together that the effect is of an underground city. It's surprisingly light too – the street-roofs are whitewashed inside & have skylights. It's not slums tho – there are stores & shops selling everything, including expensive jewelry. . .

I think I told you about eating spaghetti at a Chinese restaurant in Trinidad with chopsticks a long time ago, didn't I? On my first trip to Trinidad? It was the first time I'd ever used chopsticks. Well, the other day I was glad I'd learned how. I landed at a field in Italy just in time for lunch at the Transient officers' mess, but found the mess had nothing but plates to eat with – no knives, forks, etc, & I hadn't brought my mess kit (too much of an office soldier). So I got a piece of board & whittled myself out a pair of chopsticks & ate lunch with them. I'd certainly never expected the ability of use chopsticks might ever mean the difference between eating & not eating!

24/12/43 - *Christmas letter (3 entries in one letter -- homesick?)*

After a long dry spell I'm suddenly deluged with mail – in the last three days I've gotten sixteen letters from just about every one – nineteen, really, because the chain letter is really four letters. . .

Interested to hear about Joe & Doris' baby; Wayne mentioned it too. I don't like this business of naming children after their parents though. After all, the main function of a name is to differentiate its holder from other people, and if you go giving everyone the same name you don't achieve that function. Even in our family having two "Al's" has been confusing at times, even though the names are actually different . . .

I opened the Christmas packages that came several weeks ago, but didn't touch them otherwise. I'm just now eating my first piece of the fruit cake you sent me . . . Thank you all. And I hope you like the few things I sent . . .

It's not a white Christmas here, except in the mountains, & not having flown over any mountains in the last few days I'm not even sure it's still white there. It's cold tho – for me. This morning I added long woolen drawers to the woolen undershirt I'd already been wearing for some time. It's not really so awfully cold – it's above freezing – but all the buildings are masonry and unheated, & I never get really warmed up, except when I fly. . .

25/12/43 – 8:30 a.m.

Well, Merry Christmas to all of you. . .

It's a cold morning; I'm going to wear my fur-lined jacket to the office.

2:30 p.m.

I've just been relaxing after a big dinner–fresh vegetable soup, salad, turkey, potatoes, sweet potatoes, cranberries, fresh butter, cake, tangerines, almonds, dates, & apple pie a la mode. Quite a feed. The dates were on the table in clusters–a branch about a foot long with a couple of dozen dates on it–and we had flowers on the table & everything.

73 - Aldro

Note new address; "MAAF" stands for "Mediterranean Allied Air Forces"

The letter below says that a wave washed into their mess hall, so he was obviously on the coast. Casablanca & Rabat, Morocco are on the Atlantic and Tunis is on the Mediterranean.

5/1/44 –

It's been pretty cold these days; we had a cold & windy spell beginning New Year's Eve and now it's beginning to come back. I'm wearing wool underwear with two woolen undershirts, wool shirt, sweater, leather jacket, & trench coat with heavy lining -- about three layers more than I usually wear at home when it's 50 degrees colder than here. It hasn't been below 32 degrees yet, you know. New Year's Day we were flooded out of our mess hall -- a big wave slammed against the building, broke every window on that side, came in and threw all the tables across the dining room -- broke some of them too. We still had a good dinner, but it was delayed. We're using the second floor of that building now. . .

I've decided not to get any Kairouan rugs – they're too expensive, & not really pretty. . .

The book "Reader's Digest Books" that Conchita sent me is a collection of 14 of the book condensations that have appeared in the Reader's Digest –- very good. One of them is "Dry Guillotine," a story about the penal colonies in French Guiana. I've flown over St. Laurent, the main colony on the mainland, & Devil's Island, the better-known but smaller one off the coast of French Guiana – I found the story very interesting. . .

I stopped in at Pantelleria not so long ago – quite a place. It's all volcanic rock, but it's all terraced & just about every square foot is under cultivation. Grapes & raisins are the main livelihood of the island. . .

My Arab laundryman just brought back my laundry, & also, for me & another of his customers in the next room, a plate of fancy Arab cookies. They're a kind of pastry, made of very coarse flour, filled with dates & with a couple of almonds in the shells stuck in the top, & the whole works apparently fried in deep fat. We had to accept them & thank him, of course, but we'll probably throw them out. The French are noted for lack of sanitation, & the Arabs are worse, if possible. I munched on one while he was here, so as not to hurt his feelings – they're not bad tasting. . .

It's raining & lightning outside, & I'm sitting here in my fur-lined jacket & fur-lined boots. A few minutes ago I saw a mosquito – apparently they never completely disappear. . .

By the way – something I'd find very useful if you could send it to me is a padlock – a good one – and a good strong hasp. Right now I have a room with a lock & a key, but the chances are about 50-50 that my next place will not have a key. And it's definitely necessary to keep the room locked. . . the Arabs will steal anything any time. They're very successful burglars – don't care if the occupant of a room is in or not. Most of the thefts I know of have been at night when the people concerned have been sleeping right in their rooms. It's happened on two occasions in my building. Clothes, luggage, money, shoes – anything they can get their hands on. One of the officers nearly caught one the other day – took a shot at him but missed. I sleep with my door locked and my gun under my pillow, and keep the room locked all day except long enough for the prisoner to clean up in the morning. Kind of a dirty war – instead of shooting Germans I'm keeping my gun handy for "friendly" Arabs. In a way you can hardly blame them for stealing – the weather's cold, clothes are scarce and expensive, there's not much chance for them to earn money – I'd be tempted too, in those circumstances.

73

Aldro

Note: Apparently, he changed his mind about not getting a Kairouan rug because I have what probably is one. It's only about 15 inches square though. It's primarily green with a red square in the middle – as he said, not really pretty. I also have a handgun I found in Mom's headboard and suspect it's the one he kept under his pillow.

15/1/44 – *First mention of Mt. Vesuvius, which erupted March 17, 1944*

Got back yesterday from a several-days' trip – very uneventful, except I caught a cold. Those people over where I was have heat in their quarters & offices & I wasn't used to it & caught a cold. It's not bad tho.

I just finished off another roll of film – haven't sent it in tho. It has some pictures of Mt. Vesuvius from this last trip.

2/2/44
Dear folks –

The sleeping bag came the other day – I've slept in it 3 nights now & it really does the job. I have 5 blankets and a sheet inside it, besides two pillows, & the quilted lining is warm enough so I don't have to spread my trench coat out on top to keep warm, as I was doing with my improvised bag.

Haven't been to Naples in quite some time – haven't had a chance to get any cameos. I did see some small ones in the local Kasbah, but they weren't much. Did I tell you about my latest Kasbah trip? One of the officers I was with ran into a small boy – about 10 years old, I guess – whom he'd struck up an acquaintance with outside the Kasbah. This kid worked in a filigree jewelry shop, & he took us into the work room & we saw the stuff being made – very interesting.

Time to get to bed if I'm to get my customary 10 hours' sleep.

73

Aldro

CHAPTER 7
Italy - 1944

The following is from a summary of his career that Dad wrote in 1956 at his mother's request.

Around February I went to Caserta, Italy, just north of Naples. That's where our headquarters was in a palace. It was a tremendous building, shaped like four squares put together to form one big square. According to the Encyclopedia Americana, it's "one of the finest in Europe, a large and richly decorated structure commenced in 1752 by Charles III of Spain and designed by Vanvitelli." Its south side, according to the encyclopedia, is 830 feet long -- I'd have estimated more than that. There was no central heating, and each room had one or more fireplaces. I once started counting chimneys, from the outside, but got lost after a hundred or so. Plumbing was almost non—existent. The place had a chapel and theater -- both elaborately decorated — and my office happened to be in the throne room, which had paintings and mosaics all over the walls and ceiling. I normally wore a sheeplined flying suit in the office to keep warm. We didn't live in the palace though. We had a tent camp a couple of miles away -- not that far I guess. Some of us -- not all-- had kerosene heaters in our tents. The heater kept the tent fairly warm, but it couldn't be left unattended and I always turned it off before I went to sleep. I slept on a canvas cot in a sleeping bag, wrapped up in 7 blankets and with my heavy trench coat on top of that. I've never been as cold anywhere as I was there, despite the fact that the temperature never dropped below freezing.

Backtracking a little, when I first went to Africa, I was in the headquarters of the Twelfth Air Force. Around January 1944 we reorganized – we established the Fifteenth Air Force, with its headquarters in Bari, Italy, Headquarters Mediterranean Allied Air

Forces, located at Caserta, and the Twelfth Air Force, somewhat smaller than it had been, with its new headquarters in Foggia, Italy – it happened to be in a five-sided building which, somehow, never came to be called the Pentagon.

After a couple of months or so at Caserta I went over to the Fifteenth Air Force at Bari, and that is where I had the most interesting job. General Twining, now Chief of Staff of the Air Force, was commander of the Fifteenth at the time. Krisberg, incidentally, was at 12th AF in Foggia – he had a room with a grand piano in it – and before that he'd been at La Marsa, in Africa, and before that in San Juan, where we'd first known each other.

The things I remember most about Italy – aside from the dirt, the destruction, and the misery of the people during the war – was the extent to which the classical forms of art were a part of the lives of the people. Opera houses were about as common as movie theaters, if not more so; there was one just around the corner from my hotel, and I always intended to go there and see what an Italian opera was like, but never did. There were a lot of oil portrait painters around, and water color and crayon artists and a lot of people who made desk ornaments and other things out of marble. Most buildings were of stone. And out in the country, everywhere you'd see primitive water pumps, powered by an ox or a miserable-looking, scrawny horse or mule, tied to a horizontal wheel and walking around in a circle.

Letters from Italy

None of Dad's letters say exactly when he was transferred from North Africa to Italy, but the preceding summary says it was "around February." Maybe the deluge of mail he mentions on Feb. 10 was due to the move.

7/2/44

I now have only two assistants, & they're both named Smith. It's not as confusing as it might be tho, because neither of them is here. They're both away on special jobs, leaving me alone with all the work.

10/2/44

Day before yesterday I got such a deluge of mail I've barely finished reading it – 10 letters from Conchita, dated as late as Jan 28, yours of Jan 31 & 27, & some other stuff. Your Jan 31 letter took only 8 days to get here – that's a record. . . Got a letter from Margaret today telling me about the elevator strike – 11 floors! The last time I stayed overnite in Algiers I had a 5th-floor hotel room with no elevator, & had to carry my luggage up & thought it was pretty bad. The other day I got a letter from Mautz, one of the old Monmouth gang. He's in China now. His letter took 8 days to get here.

I finished another roll of Kodachrome a few days ago & sent it in – you'll get it. It's got some pictures of Mt. Vesuvius on it. Too bad you let the projector scare you out of trying to set it up for so long – it's very easy. In hot weather, tho, you may find that the focusing barrel will stick. I used to keep that part in the refrigerator to avoid that.

11/2/44

Been having an interesting but aggravating time the last few days – testing and tuning up a new transmitter that doesn't want to be tuned. I don't often get a chance to get inside a transmitter any more, & I've been making the most of this chance while I have it.

6/3/44

Quite a while since I've written you, I guess. I've been pretty busy; went on a several-days joy-ride that included a day at Casablanca and stops at other places in French Morocco, & lots of things. . .

The other day I had my first Coca Cola since coming over here. They're making it over here now, but it's rationed – 3 bottles a week.

I sent some boxes to Madison addressed to Conchita – they're mostly some leather hassocks I got in Morocco. . . You may as well open them all – the hassocks are rolled up pretty tight & it will be best to get them unrolled. I don't know when Conchita will be going to Madison, but expect it may not be likely to be long after those packages get there. She keeps talking about my coming home about May or so – I don't know where she got the idea & I can't convince her that that's extremely unlikely. Combat crews get sent home after completing a certain number of combat missions, but staff officers stay on forever. Which is as it should be, I guess.

27/3/44 *A transfer, and Mount Vesuvius eruption*

Dear folks –

A transfer that I've been working on ever since I hit this theater finally came thru. . .

Pretty nice here – aside from the fact that the job is exactly what I've wanted to do, the living accommodations are comfortable (I have a bed with an inner-spring mattress & sheets), the office well-laid-out & comfortable, & the food fair. The food's not as good as back at MAAF tho – that would be too much to expect. However, I did have imported American ice cream last nite, & was it good! Three big helpings.

I suppose the papers have mentioned Mt. Vesuvius' eruption. I have some pretty good pictures I took of it, both from the air & from the ground. I hope they're good anyway – I haven't had them developed yet. If they're not it's too bad; I'm not going to stay in Italy to wait for the next chance; the last one was 38 years ago.

We're having the same kind of a winter as you – the thermometer hasn't been below freezing all winter, but now that the equinox has gone by & spring is supposedly here, there have been snowflakes in the air twice. They melted as soon as they touched the ground, but they were in the air.

1/4/44.

Changing to daylight saving time tomorrow – that will be 8 hours earlier than CST. "CST" is an almost unknown term to me now – I've been thinking so long in terms of Greenwich time and "zone + 4", "zone – 1", or "zone Q", "zone 1", "zone Z" etc. You're in zone "plus six" time & I'm in zone "minus two" time, effective tomorrow. Or, expressed the other way, you're on "S" time & I'm on "A" time; tomorrow I'll be on "B" time. Conchita's on "Q" time. Understand it? She does – she had to learn it.

21/6/44 *(Grandma did save the June 6, 1944 edition of the Madison paper, which reports on the events of the same day!)*

Got your 6/6 V-mail yesterday – the invasion seems to have stirred up quite a storm in the States. Did you save the newspaper that announced it? I hear the newspapers in some towns printed no ads that day – nothing but invasion news. . .

Been out flying "Snoozin Susan" today; also had my picture taken in the cockpit, which I'll send you.

73--

Aldro

11/7/44

Got a V-mail today from Conchita, written July 15 on a train between Miami & New York. . .

I'd like to see the Vesuvius pictures – how about sending me some of them? I'm at a disadvantage here; I can't have Eastman return those pictures to me; they have to be delivered in the States. But there's nothing to prevent you from sending them to me. And incidentally, how about sending me some color film? Get "Daylight Kodachrome (<u>not</u> Type "A" Kodachrome) in 18-exposure 35-

millimeter rolls . . . They cost about $2.50 a roll, & Conchita will pay for them.

Been out flying for a while today. I was out all day yesterday with a B-17 showing a party of brass hats around. Today I took them to their next stop, & then had to hurry back to meet two more parties that had descended on me. You can tell this is the season of good flying weather across the Atlantic – the Washington & London junketeers have been holding a continual convention in my office. In the last two weeks I've had 13 visiting firemen to take care of. . . I told Conchita to send me a cap marked "GUIDE" . . .

I checked off in a B-26 & a B-24 last week, so now I'm qualified to fly any airplane in the 15th Air Force except one fighter, & I don't care if I ever fly that or not. I could have flown these two ships before but had no particular inclination to – the B-17 & B-25 satisfy me pretty well. And still do, altho I didn't find as many objectionable features in the B-26 & B-24 as I'd heard about.

Got a questionnaire from the War Manpower Commission today. They want to know what kind of war work I might be available for!

73 –

Aldro

29/7/44

Dear folks

I imagine you know by this time that I'm coming home soon. Just how soon I don't know. Right now I'm at a rest camp (and wait until I can tell you about it!) & when I get back to HQ my orders should be about ready. That doesn't mean I'll be home right away because I may go by boat. . . I'm going to try to fly. . .

Lots of sightseeing to do, both historical & religious. I've taken a lot of pictures, but they don't compare with the scenes themselves. I was in one church, for instance, that was the most beautiful I've ever seen (also the biggest), & I didn't take any pictures at all in there because I couldn't have begun to do the place justice.

This is the first rest camp I've been to and it's really luxurious. It's not a camp at all, but a fancy hotel. You can sleep every morning until 9:15 and still get breakfast, & when you get up, you find today's Stars and Stripes under your door. Had pancakes, syrup, & butter this morning (everyone in Italy has butter except HQ 15th AF!). A block away is the Officers' Red Cross Club, with a nice big light cool reading room, snack bar, & the works. I've run into a lot of people I know here.

73 –

Aldro

30/7/44 *(his last letter from Italy was to Mom in Madison)*

Darling,

Haven't done much today – stayed in bed until 9 AM (they serve breakfast until 9:30), went for a short sightseeing trip, spent some time reading and relaxing at the Red Cross and then did the same thing this afternoon. My routine was slightly disturbed by the news that the airplane that's to take me & a bunch of other people back to HQ tomorrow won't be able to land at the prearranged field, but I've finally gotten that fixed up. Wouldn't do for me to be delayed getting back when I've probably got a stack of letters from you waiting for me and my going - home orders (I hope) and everything. . .

I got you some silk stockings yesterday, just a couple of pair because I don't think they're much good. The salesgirl said they were size 4 ½ , which is just right, except I didn't know if their sizes are the same as ours. So, I had her measure the length of the foot, which was 25 centimeters; yours is 23 centimeters long, so I guess they'll fit OK. The size similarity surprised me.

Best love, honey –

Al

EPILOGUE

A Short Biography

Aldro Immanuel Lingard was born May 24, 1917 in Mt. Horeb, Wisconsin, a small town near Madison. His family had no car or bathroom. In the space of a lifetime, he went from reading by the light of kerosene lamps to directing the development of advanced aircraft instrumentation systems.

He grew up during the Depression, graduated from the University of Wisconsin, and later went on to earn Masters and Doctorate degrees at MIT.

"Wise men say only fools rush in," but Mom and Dad rushed in, getting married just 24 hours after deciding to do so, and they lived happily ever after until Dad died at the age of 66.

He and Mom were model parents, and I can honestly say that I never saw them argue. They were always loving, generous and supportive, even when we were taking off to live thousands of miles away. In fact, Dad may have sown the seeds for our adventures when he would bundle us up in our station wagon to drive across the United States or, when stationed in Germany, to a dozen different European countries.

During his 30-year career in the Air Force, Dad never talked about his work, presumably because it was classified. But his lifelong friend and fellow Air Force officer Carl Walter later informed us that "he dealt with new air navigation devices, autopilots, new VHF/UHF radios, identification beacons, etc. In the military he had what he enjoyed most . . . he kept technologically challenged, he kept flying, he gave his family an active life, and he was in at the birth of many exciting changes."

Made in the USA
Las Vegas, NV
14 November 2026

12930519-8db5-4c19-bbca-e2642f8fa742R02